GROUND MOVEMENTS INDUCED BY SHIELD TUNNELLING IN NON-COHESIVE SOILS

ADVANCES IN GEOTECHNICAL ENGINEERING AND TUNNELLING

General editor:

D. Kolymbas

University of Innsbruck, Institute of Geotechnics and Tunnel Engineering

In the same series (A.A.BALKEMA):

1. D. Kolymbas (2000), *Introduction to hypoplasticity*, 104 pages, ISBN 90 5809 306 9

2. W. Fellin (2000), *Rütteldruckverdichtung als plastodynamisches Problem, (Deep vibration compaction as a plastodynamic problem)*, 344 pages, ISBN 90 5809 315 8

3. D. Kolymbas & W. Fellin (2000), *Compaction of soils, granulates and pow-ders - International workshop on compaction of soils, granulates, powders*, Innsbruck, 28-29 February 2000, 344 pages, ISBN 90 5809 318 2

In the same series (LOGOS):

4. C. Bliem (2001), *3D Finite Element Berechnungen im Tunnelbau, (3D finite element calculations in tunnelling)*, 220 pages, ISBN 3-89722-750-9

5. D. Kolymbas (General editor), *Tunnelling Mechanics, Eurosummer-school, Innsbruck, 2001*, 403 pages, ISBN 3-89722-873-4

6. M. Fiedler (2001), *Nichtlineare Berechnung von Plattenfundamenten (Nonlinear Analysis of Mat Foundations)*, 163 pages, ISBN 3-8325-0031-6
7. W. Fellin (2003), *Geotechnik - Lernen mit Beispielen*, 230 pages, ISBN 3-8325- 0147-9
8. D. Kolymbas, ed. (2003), *Rational Tunnelling, Summerschool, Innsbruck 2003*, 428 pages, ISBN 3-8325-0350-1
9. D. Kolymbas, ed. (2004), *Fractals in Geotechnical Engineering, Exploratory Workshop, Innsbruck, 2003*, 174 pages, ISBN 3-8325-0583-0
10. P. Tanseng (2005), *Implementations of Hypoplasticity and Simulations of Geotechnical Problems*, in print.
11. A. Laudhan (2005), *An Approach to 1g Modelling in Geotechnical Engineering*, 204 pages, ISBN 3-8325-1072-9
12. L. Prinz von Baden (2005), *Alpine Bauweisen und Gefahrenmanagement (Alpine Construction Methods and Risk Management)*, 243 pages, ISBN 3-8325-0935-6
13. D. Kolymbas & A. Laudahn, eds.(2005), *Rational Tunnelling, 2nd Summerschool, Innsbruck, 2005*, 304 pages, ISBN 3-8325-1012-5
14. T. Weifner (2006), *Review and Extensions of Hypoplastic Equations*, in print
15. M. Mähr (2006), *This book.*

Ground movements induced by shield tunnelling in non-cohesive soils

Markus Mähr

University of Innsbruck, Institute of Geotechnics and Tunnelling

E-mail: markus.maehr@gmail.com

Homepage: http://geotechnik.uibk.ac.at/

The first three volumes have been published by Balkema
and can be ordered from:

A.A. Balkema Publishers
P.O.Box 1675
NL-3000 BR Rotterdam
e-mail: orders@swets.nl
website: www.balkema.nl

Bibliographic information published by the Deutsche Nationalbibliothek

The Deutsche Nationalbibliothek lists this publication in the Deutsche Nationalbibliografie; detailed bibliographic data are available in the Internet at http://dnb.d-nb.de .

ISBN 3-8325-1361-2
ISSN 1566-6182

Logos Verlag Berlin
Comeniushof, Gubener Str. 47,
10243 Berlin
Tel.: +49 030 42 85 10 90
Fax: +49 030 42 85 10 92
INTERNET: http://www.logos-verlag.de

Contents

Vorwort des Verfassers

Das ständige Wachstum der Ballungsräume und die zunehmende Mobilität der Menschen führt zu einem steigenden Bedarf an Transportmöglichkeiten. In stark bebauten Gebieten kann dieser Bedarf durch unterirdische Verkehrswege platzsparend gestillt werden. Aus praktischen Gründen verlaufen die Tunnel meist knapp unter der Oberfläche, was beim Bau zu Bodenbewegungen führt, die die umgebenden Gebäude gefährden.

Die Prognose der Bodenbewegungen infolge Tunnelbaus ist ein aktuelles und lebendiges Gebiet der Forschung. Mein Ziel war es, mittels moderner numerischer und experimenteller Methoden das Verständnis der Mechanik des Tunnelbaus zu vertiefen und darauf aufbauend die Prognoseverfahren für Setzungen zu verbessern. Schließlich wurde auch eine Idee für eine neue Methode zur Verhinderung von Setzungen geboren.

Viele haben mir dabei geholfen diese Arbeit fertigzustellen. Sei es direkt, indem sie mir Anregungen gegeben, Versuche durchgeführt oder Versuchsapparate gebaut haben, oder indirekt, indem sie mir zugehört haben und mich meine Gedanken ordnen ließen.

An erster Stelle möchte ich meinem Doktorvater Prof. Dimitrios Kolymbas danken. Er hatte nicht nur die ursprüngliche Idee zu dieser Arbeit, sondern ließ mich auch gewähren, als ich dieses Thema dem von ihm vorgeschlagenen vorzog. Gegen Ende der Arbeit half er mir sehr durch seine Korrekturen und Anregungen deren Qualität zu verbessern.

Weiters möchte ich Prof. Ivo Herle und Dr. Wolfgang Fellin danken. Ersterer hat die mühevolle Arbeit des Koreferenten auf sich genommen. Beide aber waren mir sehr wertvolle Ratgeber und Vorbilder, die immer ein offenes Ohr für meine Fragen hatten.

Einen wesentlichen Beitrag lieferten auch meine Diplomanden. Heiner Fromm wirkte an der Entwicklung des Versuchsapparates mit. Jana Bochert und Andrew Ross führten zahlreiche Versuche durch.

Dank gebührt auch Peter Deseife für seine Hilfe bei der Entwicklung des Ringspaltgeräts und Herbert Martini, der es verstand unsere Ideen in die Realität umzusetzen.

Allen Mitgliedern des Geotechnik Instituts möchte ich für Ihre Unterstützung und für das sehr angenehme und familiäre Arbeitsklima danken. Ich habe die

Zeit am Institut sehr genossen und werde mich gerne daran erinnern.

Schließlich möchte ich mich noch bei meiner Lebensgefährtin Karin Pürmair für die längjährige Rücksichtnahme bedanken. Die größte Freude hat sie mir jedoch mit unsere Tochter Pia bereitet, die in dieser Zeit geboren wurde.

Röthis im Juli 2005
Markus Mähr

Chapter 1

Introduction

1.1 Settlements due to tunnelling

Rapid growth in urban development has resulted in an increased demand of tunnels. For obvious practical reasons such as accessibility, serviceability and economy, tunnels are often constructed at shallow depths through soft soil or weak rock [88].

About 80% of soft ground tunnels recently completed or currently under construction are excavated using tunnel boring machines (TBM) [26]. In this category, the earth pressure balance or slurry TBM with controlled face pressure are the most widely used ones, as they offer superior ground control capabilities. Recent technical developments in shield tunnelling technique caused a decrease of ground deformation due to tunnelling, however the deformations of the ground surface are still noteworthy.

The causes of settlements are various and interact with each other. The interaction between TBM and the ground belongs to the most complex soil-structure interaction problems in geotechnical engineering. Attempts are made to model this complexity by analytical, laboratory or numerical modelling. Although these techniques are still of limited use for practical purposes, they help to understand the key features of these tunnelling techniques.

1.2 Objective and overview

The objective of this thesis is to investigate soil behaviour due to shield tunnelling. Special focus is put on effects due to the volumetric behaviour of soil, which are neglected in the majority of former research. This investigation is done by means of small scale laboratory models and numerical analyses.

The thesis starts with a literature review in *Chapter 2*, that presents the currently used design approaches for predicting tunnel induced soil movement and summarises model tests, analytical methods and numerical studies, both 2D and 3D.

A comparison of FE analyses with hypoplasticity and elasto-plastic material model with Mohr-Coulomb yield-criterion in *Chapter 3* shows that Hypoplasticity is one of the few constitutive models, which is able to reproduce correctly volumetric soil behaviour for loading-unloading paths that occur in shield tunnelling. Therefore Hypoplasticity with intergranular strain is used to simulate model tests. To ensure accuracy and speed of the calculation a second order method was adopted for the time integration of the state variables and the determination of the Jacobian matrix.

In *Chapter 4* a laboratory model (scale 1:50) to analyse the soil behaviour due to shield tunnelling and tail gap grouting is presented. The influence of the initial density on settlements is pointed out and strain localisation (arching and shear planes) is investigated.

In *Chapter 5* FE simulations of the model tests are presented. The results emphasise the appropriateness of Hypoplasticity for shield tunnelling simulations.

A concept to avoid settlements is proposed in *Chapter 6*. An injection scheme for compaction grouting is presented and its efficiency is tested by means of FE simulations.

The own model tests and their FE simulations show that the known analytical solutions to predict surface settlements lack the consideration of volumetric soil behaviour. Thus, in *Chapter 7* a new analytical method to calculate the volume of the settlement trough is presented that takes volumetric soil behaviour into account.

Finally conclusions are collected in *Chapter 8* and an outlook is given.

Chapter 2

State of the art

2.1 Shield tunnelling

Fig. 2.1: Earth Pressure Balance (EPB) shield: (1) working face, (2) cutting wheel, (3) excavation chamber, (4) pressure wall, (5) tunnelling jacks, (6) screw conveyer, (7) lining segment erector, (8) tunnel lining with lining segments [4].

Shield tunnelling is beside conventional tunnelling methods such as NATM a well established method which allows for tunnel advances in a wide range of soils and difficult conditions such as high ground water pressures, soft soils or small cover.

An important question related to shield tunnelling is how to stabilise and support the face. The face can be supported in different ways: mechanical means, compressed air, fluid or by the excavated soil itself.

In soft ground the mechanical support of the face is not suitable and very risky, especially below the ground water level. The compressed air shield was the first approach used in soft ground, followed by the slurry shield, where the face is supported by a fluid (usually water and additives such as bentonite). Lastly the Earth Pressure Balance shield (EPB) was introduced. Here the excavated soil material itself supports the face. The selection of the system depends on the geological and hydrological situation.

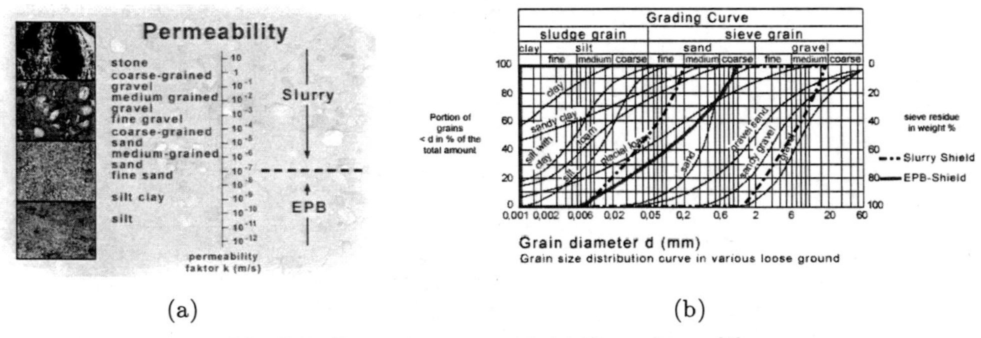

(a) (b)

Fig. 2.2: Operative range of shield machines [5]

Despite recent advances in tunnelling technology, tunnelling-induced ground movements and thus settlements of the surface cannot be completely prevented. Their origins are [9]:

- Ground loss ahead of the tunnel face due to deformation of the ground ahead of the face (termed **face loss**),
- shearing of soil at the shield-soil interface,
- over-excavation (termed **shield loss**) due to
  - pitching angle (poor workmanship),
  - curvilinear driving path,
  - conicity of the shield,
- tail gap and tail gap grouting (termed **tail loss**),
- deformation of the lining,

– long term consolidation due to excess porewater pressure dissipation and changes in ground water hydraulic conditions.

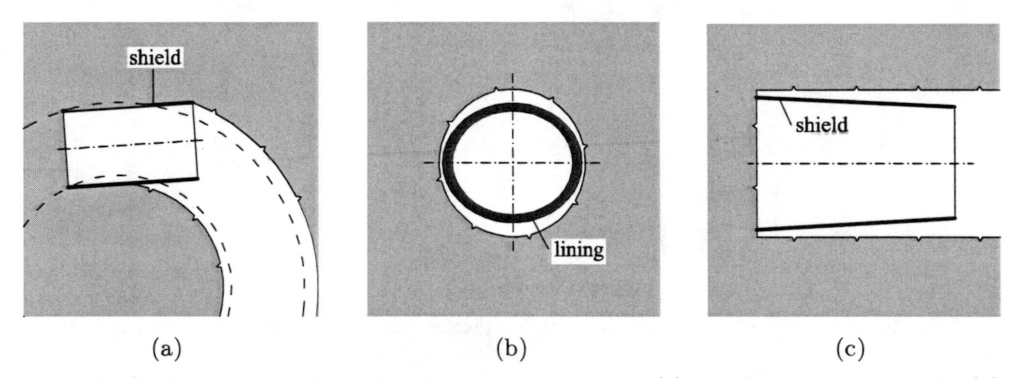

Fig. 2.3: Settlements may be induced amongst others by (a) curvilinear driving path, (b) deformation of the lining and (c) conicity of the shield.

The amounts of settlement induced by these factors depend on the applied technology as well as on hydrological and geological circumstances. Several authors apportion the settlements sources [24, 45, 56].

Ground	Ahead of shield (%)	Over the shield (%)	Tail of shield (%)
Sand above water table	30-50	10-20	60-80
Sand below water table	0-25	0-20	50-75
Stiff clay	30-60	0-20	50-75
Silt and soft clay	0-25	10-40	30-50

Tab. 2.1: Typical partition of the total settlement to the various types of ground loss at a shield-driven tunnel [24].

The settlement due to the soil deformation into the face can be controlled by adjusting the support pressure upon the face or the discharge of extracted soil. The closure of the tail gap is counteracted by grouting cement mortar into the gap (see Fig. 2.4). The replacement of ground with mortar is not strictly simultaneous with excavation, so that this process can be conceived rather as a closure of the tail gap followed (more or less soon) by an expansion of the gap by grouting. It is, however, observed that this closure-expansion cycle does not annihilate the surface settlement [45]. This is so even if the volume of the grout substantially exceeds the volume of the tail gap.

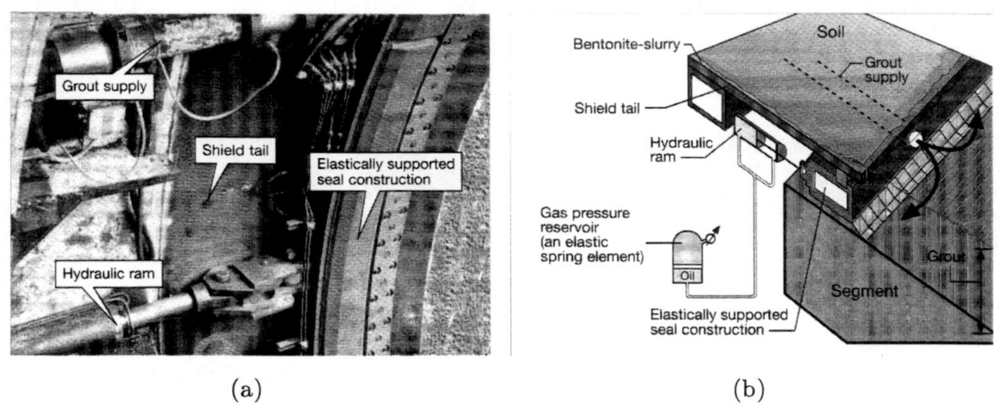

(a) (b)

Fig. 2.4: Grouting of the tail gap with a movable elastically supported seal [44]

2.2 Case histories

Although a large number of tunnels has been built up to now with shield machines, only few of them have been accompanied by extensive measurements of the ground movement and only a part of these measurements have been published. Two case histories are presented, to give an idea of the problems related to ground movement around shield driven tunnels.

2.2.1 Nagata

Soil conditions and tunnel geometry
The first case study published by Hashimoto et al. [37] deals with a tunnel construction site in Nagata, Japan. The underground consists of sensitive clay with an unconfined compressive strength of $q_u = 70$ kPa and an index of liquidity of $I_L = 0.9$. The tunnel, whose diameter is 6.93 m, was excavated with an EPB-shield, using simultaneous back-fill grouting with two-component mortar.

Observed displacements
Using settlement gauges and inclinometers the displacements of the ground in a cross section around the tunnel shown in Fig. (2.6) were measured. The deformation in longitudinal section can be seen in Fig. (2.5). One can clearly see the cyclic movement of the tunnel. In front of the face the soil moves toward the tunnel centre, over the shield the soil moves outward and behind the tail the soil moves inward again.

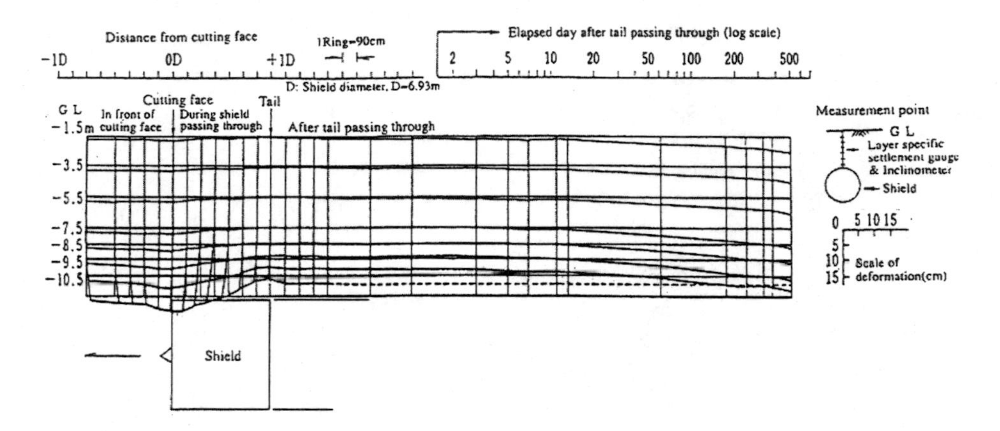

Fig. 2.5: Ground movements due to shield tunnelling in soft clay (Nagata), longitudinal section [43]

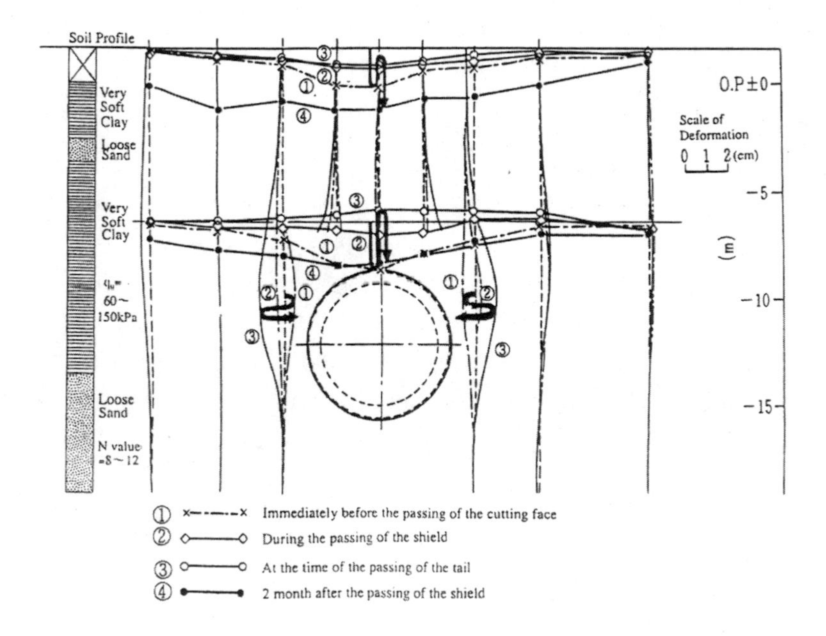

Fig. 2.6: Ground movements due to shield tunnelling in soft clay (Nagata) [43]

The authors point out that the subsequent settlements, which occur after the shield machine passed through, are mainly caused by consolidation of the disturbed ground. The disturbed area is limited and the disturbance is mainly compression (see Fig. 2.8). Furthermore they postulate that the larger the maximum deformation during shield passage is, the larger the subsequent deformation will be. This correlation is depicted in Fig. (2.7).

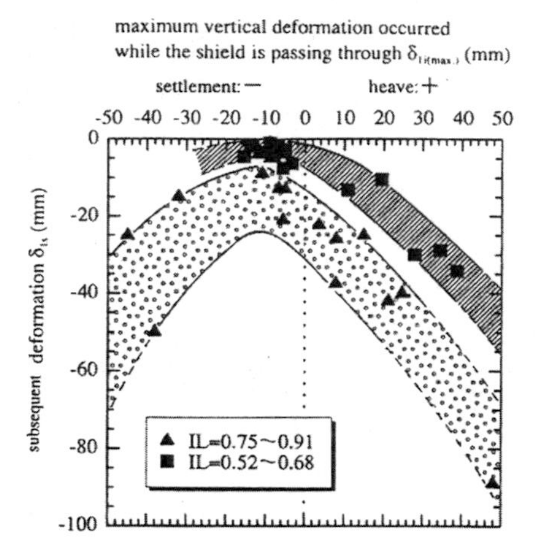

Fig. 2.7: Relation between the maximum deformation during the shield passing and the subsequent deformation at 1 m above the crown [37]

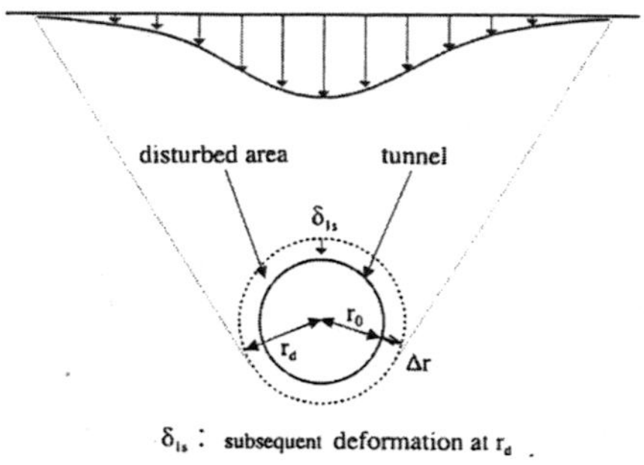

Fig. 2.8: Conceptual diagram of subsequent deformation [37]

2.2.2 Washington D.C. rapid transit system

The second case study presented here is an extensive field monitoring program of the construction of the Washington D.C. rapid transit system in its initial years (1971 and 1972). The measurements presented by Hansmire and Cording [36] are one of the rare case histories of shield tunnels in predominantly cohesionless soil.[1]

Soil conditions and tunnel geometry

The soil in the upper part of the cross section consisted of cemented sand, gravel and silty sand. The rest consisted of silty sand or silty clay. The tunnel axis was situated 14.7 m below the ground surface, the diameter was 6.4 m giving a H/D ratio of 2.3.

Method of construction

Prior to tunnel construction, dewatering was carried out using large diameter wells. The tunnel was excavated using a ripper bucket machine.

[1]Only the measurements of the first tunnel from the pair of tunnels will be presented here.

Observed displacements
It was found by Hansmire and Cording [36] that major soil movements occurred over the shield and due to the tail gap and that there was very little time-dependent movement. The ground movements that took place during the construction of the first tunnel are summarised in Fig. (2.9a). The vertical displacements are attenuated towards the ground surface. Due to loosening of soil above the crown the measured volume of the settlement trough $V_S = 1.7$ m^3/ m was smaller than the volume loss $V_L = 2.5$ m^3/ m, i.e. soil volume excavated in excess of the crosssection. The major zone of movement above the tunnel was confined to a region slightly wider than the tunnel diameter. The zone was bounded by shear failure in the soil above the sides of the tunnel (cf. Fig. 2.9b).

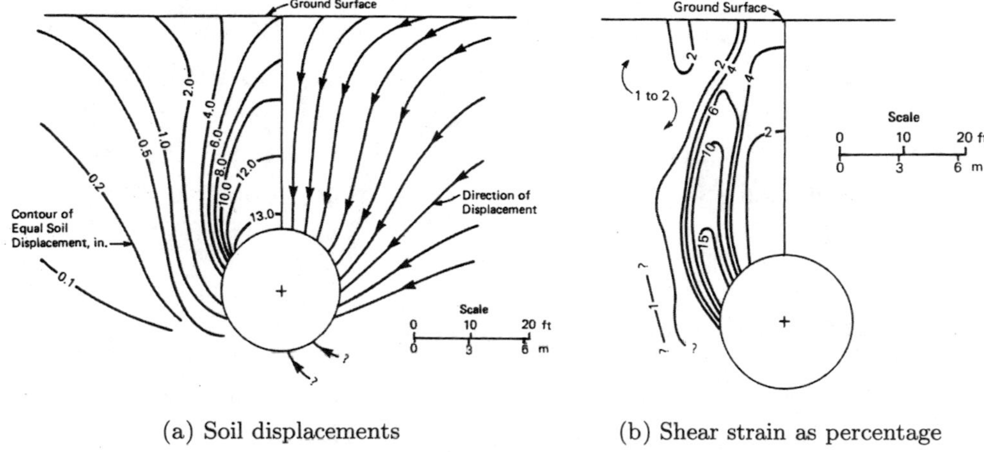

(a) Soil displacements (b) Shear strain as percentage

Fig. 2.9: Measurements of Washington D.C. rapid transit system [36].

The observed face loss (lost ground ahead of the shield) was very small, while shield loss (lost ground while shield passed the measurement section) was the largest component of lost ground (about two thirds of total ground loss). This was due to the fact that the shield was difficult to steer. As a result, a relative big pitching angle occurred, which caused ground loss. Tail loss was about one third of the total ground loss and mainly caused by lining deformation.

The pattern of ground movement was strongly influenced by volume changes in the coarse-grained soils through which the tunnel was driven.

Analysis
Rowe and Kack [87] performed a FE analysis using the Gap parameter method

(see section 2.5.2.1). The FE analysis was carried out using an elastic-plastic material model with Mohr-Coulomb yield-criterion. The resulting settlement trough was too deep and too wide. In a second trial the Young's modulus was considered to represent a secant modulus. Thus, up to a distance of 2.5 m from the tunnel contour the Young's modulus was decreased due to the large strains obtained in the first analysis. The second trial gave a sufficiently good estimation of the settlements.

2.3 Model tests

The results of field measurements and tests, as presented in the previous section, are difficult to interpret. Adverse environment may cause malfunctioning of instrumentation and they are not reproducible. It is almost impossible to correlate the results of two field tests carried out at different sites. Hence, they do only marginally contribute towards improving our understanding of the behaviour of tunnels and may not be very useful from the point of view of calibrating new design methods. Many of the limitations of field tests can be avoided if model tests are used in exploring the processes related with tunnelling [92].

2.3.1 Influence of dilatancy and contractancy on settlements

Nakai et al. [69] investigated the influence of dilatancy on soil deformation using Kaolin clay powder and Toyoura sand to carry out modified trap-door tests. They used the apparatus sketched in Fig. (2.10) with several movable blocks. 3D advancement of the shield was simulated by lowering one block after the other. The results have been compared to a 2D test using the same apparatus, where all blocks have been lowered at once. In Fig. (2.11) one can see that the surface settlements of dilatant sand is remarkable smaller than surface settlement of contractant clay. Furthermore, the difference of the settlements in the 3D simulations is more pronounced than in the 2D simulations.

2.3.2 Influence of construction process

Nomoto et al. [75] developed a miniature shield (depicted in Fig. 2.12) to simulate the shield tunnelling process in a centrifuge. Centrifuges are used

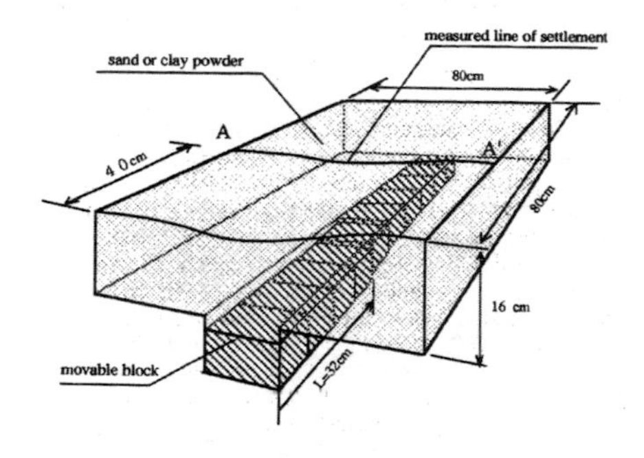

Fig. 2.10: Sketch of apparatus [69]

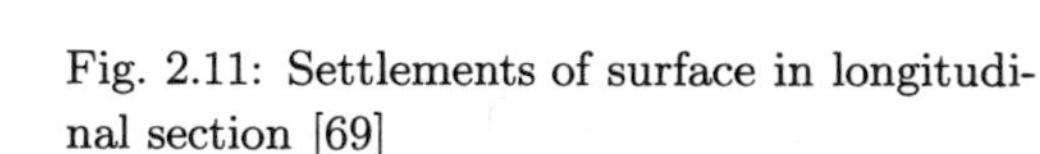

Fig. 2.11: Settlements of surface in longitudinal section [69]

to increase the body force in small scale tests, so that even in a small model the same stresses can be realized as in the prototype problem. The shield is modelled by a steel tube (diameter = 100 mm), which houses the screw conveyor with a cutter at the face (see Fig. 2.13). Behind the shield a second steel tube (diameter = 96 mm) follows, which serves as tunnel lining and on whose surface a series of load cells are inlaid (this part is called lining tube in Fig. 2.13). An outer tube (diameter = 100 mm) for simulating the tail void formation encloses the lining tube and may be pulled out. This tube is termed shield tube. With this model setup they carried out three test series:

Buried tube test: The lining is directly buried in the container[2] and the earth pressure acting on the lining tube is measured (Fig. 2.14a).

Tail void test: This test modelled the tail void formation. The shield is buried in the container and the shield tube is pulled out to create a tail void (Fig. 2.14b).

Shield test: In this test the complete process of shield tunnelling is simulated. After the shield is pushed 230 mm trough the sand, the shield tube is pulled out to create a tail void (Fig. 2.14c).

The test results clearly show that the distribution of the earth pressure acting on the tunnel is largely dominated by the construction processes (cf. Fig. 2.15). From the tests an equation for settlements in longitudinal direction is derived by Nomoto et al. [75] (see eq. 2.22, p. 26).

[2] The authors do not describe in detail how the lining is buried.

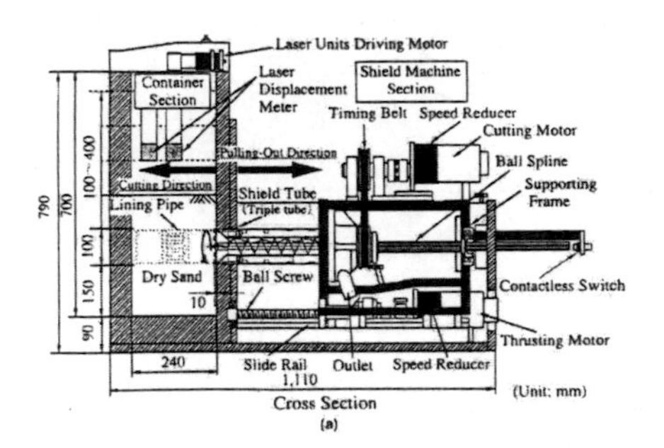

Fig. 2.12: Centrifuge model [76]

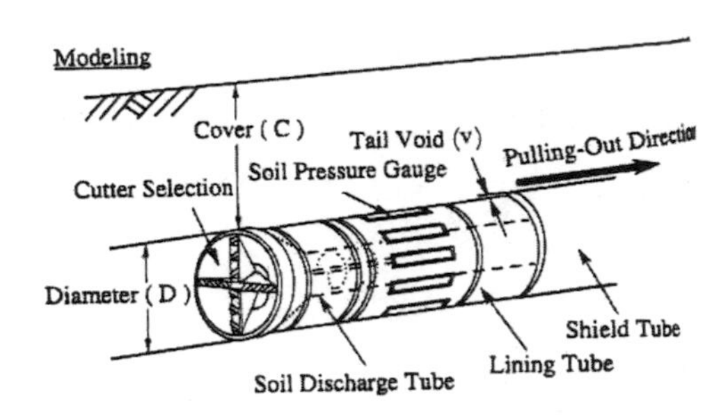

Fig. 2.13: Tail void formation process [76]

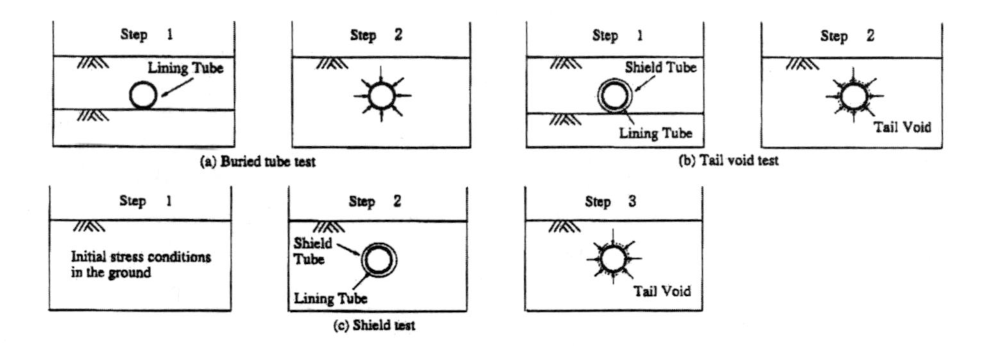

Fig. 2.14: Shield construction sequence in centrifuge tests [76]

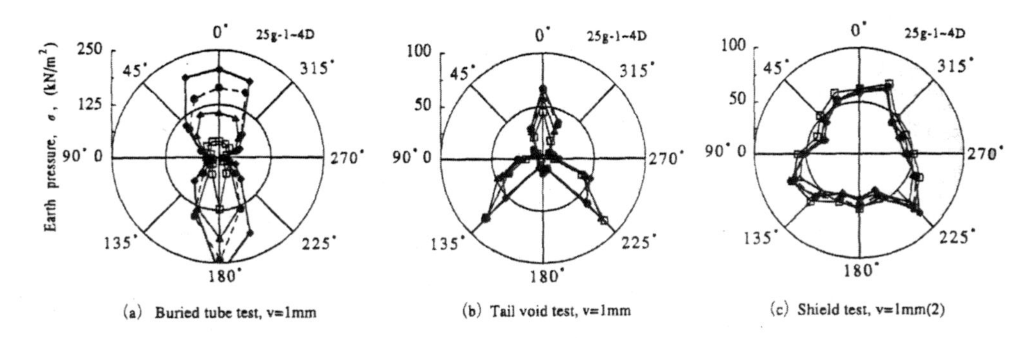

Fig. 2.15: Load cell measurement results of various tests according to Nomoto et al. [76]

2.3.3 Failure mechanism of shallow tunnels

Melix [67] investigated the stability of shallow tunnels in non-cohesive materials far from the face (2D case) and near the face (3D case). The model box was made of perspex (see Fig. 2.16b) to take photographs of the soil during the tests. In the 2D case, the tunnel was driven with a pipe using a cutter head to remove the soil. The tunnel crossed the model box. Then either a surface load was applied by a flat jack (2.16a) or the internal pressure from a rubber membrane, which modelled the lining, was reduced. In the 3D case the tunnel was driven up to the middle of the box. Then the rubber membrane was installed and a load (flat jack or reduction of internal pressure) was applied.

The results from the model tests were compared with analytical solutions to estimate the collapse load. Neither kinematic nor static or mixed methods yield good results for all tests.

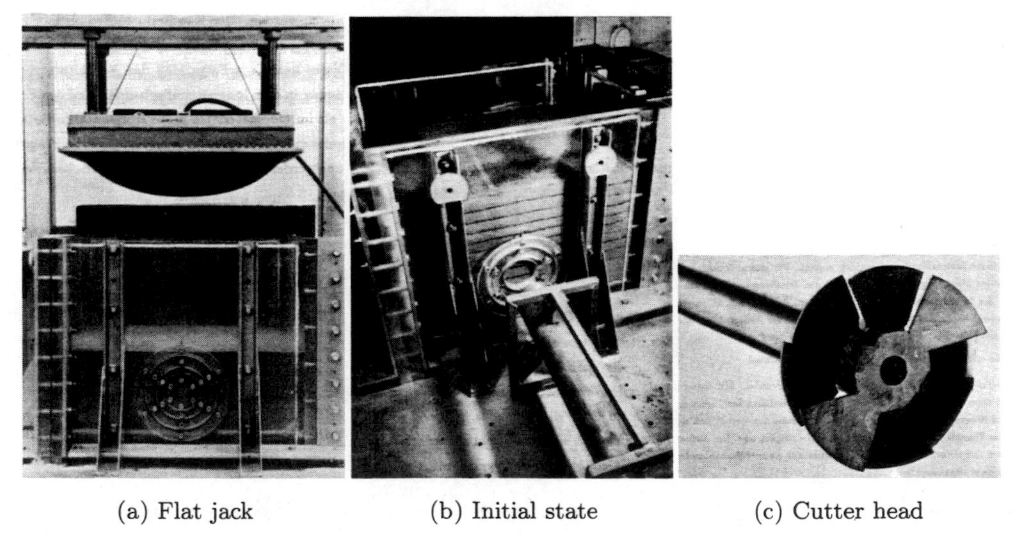

(a) Flat jack (b) Initial state (c) Cutter head

Fig. 2.16: Experimental setup for the tests of Melix [67]

In the plane strain tests Melix [67] observed two different failure mechanisms. In the first mechanism (Fig. 2.17a) two shear planes develop (1) starting from the cross section of the tunnel. They come together above the crown and form an arch. With ongoing settlements at the crown further arches develop (2,3) until the surface is reached and daylight collapse occurs.

In the second failure mechanism three blocks appear (Fig. 2.17b). The one above the crown (1) is defined by the envelope of the previously described arches. Additionally, at the sides of this block two shear planes, which start at the tunnel cross section, delimit the other blocks (2).

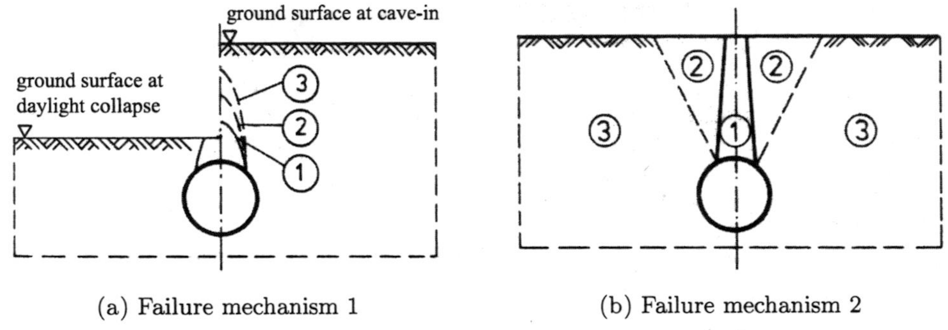

(a) Failure mechanism 1 (b) Failure mechanism 2

Fig. 2.17: Failure mechanism observed by Melix [67]

2.3.4 Crown load of shallow tunnels

König [48] developed a method to slightly reduce the diameter of the tunnel in a centrifuge test. For this purpose a mixture of bentonite, particulata silica and water is placed around the tunnel lining and defines the contour of the tunnel (Fig. 2.18a). During the centrifuge test the mixture consolidates due to the increased acceleration forces and hence the tunnel diameter decreases. The tests have been conducted to improve a method proposed by Stoffers [94] to estimate the load on a tunnel lining, which recommended the use of a reduced friction angle to calculate the tunnel load with e.g. the failure mechanisms of Therzaghi [97] or Bolton [16]. The tests proved the method of Stoffers to give good estimates of the crown load.

2.3.5 Summary

The outlined tests show that:

- dilatancy of soil has remarkable influence on the settlements,
- 3D effect is more pronounced in dilatant soil than in non-dilatant soil,
- the construction process (i.e. the load history) influences the resulting stresses acting on the lining,

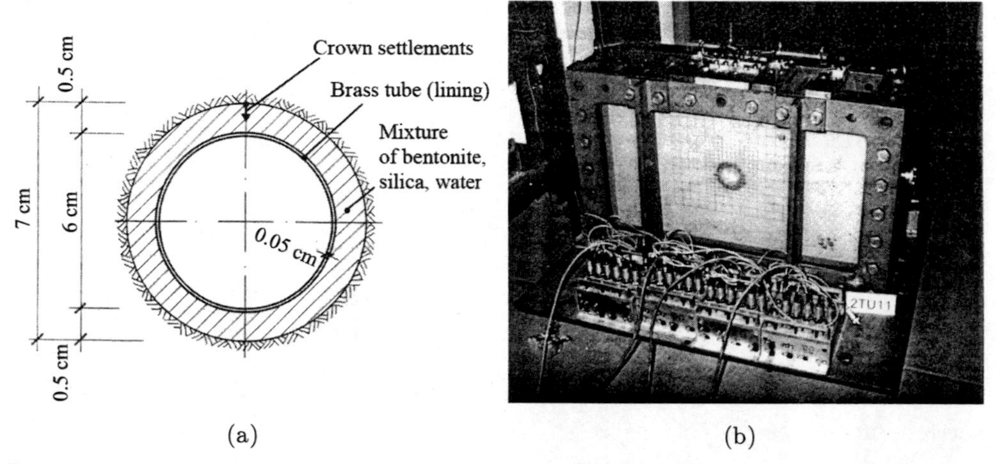

Fig. 2.18: Experimental setup for the tests of König [48]. (a) A mixture of bentonite, silica and water placed around a brass tube defines the tunnel contour. (b) Centrifuge box with tunnel in the centre and measurement devices.

- rigid body failure mechanisms dominate the deformation patterns in granular soils.

Although several different model tests have been developed to investigate soil behaviour due to tunnelling in the past, none of them investigated the effect of tail gap grouting.

2.4 Analytical approaches

Analytical approaches are still a widespread tool in tunnel engineering. This may originate from their simplicity as well as the experience many engineers collected using them. Besides the fact that they are a useful tool to control numerical calculations, they help to understand what is going on in the underground.

However, to get useful equations some assumptions have to be introduced. The analytical solutions known for the time being are mostly restricted to linear-elastic soil behaviour. Due to the known nonlinear stress-strain-relationship of soil, these assumptions are, therefore, rough approximations and one should be aware of their limited accuracy.

Since the influences on settlements, as already mentioned, are various and complex, the analytical solutions are often "improved" by empirical corrections. One should be aware, that empirical formulas do not take into account tunnel geometry, construction technique or soil conditions.

2.4.1 Surface settlements in transverse direction

2.4.1.1 Method based on LAMÉ's solution

Surface settlements may be treated by using LAMÉ's solution of the problem of a thick walled cylinder, loaded by the hydrostatic stress σ_∞.

Assuming, that the cylinder is infinitely thick, the distribution of the surface settlement u_z results to:[3]

$$u_z = \frac{u_{z,\max}}{1 + (y/H)^2} \quad ,$$

with H being the overburden, y being the horizontal distance from the tunnel axis and $u_{z,max.}$ being the surface settlement above the crown. The maximum displacement reads:

$$u_{z,\max} = \frac{\sigma_\infty - p}{2G} \cdot \frac{r_0^2}{H} \quad , \tag{2.1}$$

with p being the support pressure and G being the shear modulus.

Celestino and Ruiz [20] adopt the solution

$$u_z = \frac{u_{z,\max}}{1 + (y/a)^b} \quad , \tag{2.2}$$

where a and b are free parameters. This version allows better fitting than LAMÉ's solution.

2.4.1.2 Method of SAGASETA

Sagaseta [89] proposed a theoretical approach to this problem, based on solutions for incompressible irrotational fluid flow. The ground loss of undrained soil is considered in an infinite space, reducing the tunnel to a point sink

[3] A more detailed derivation can be found in Kolymbas [51].

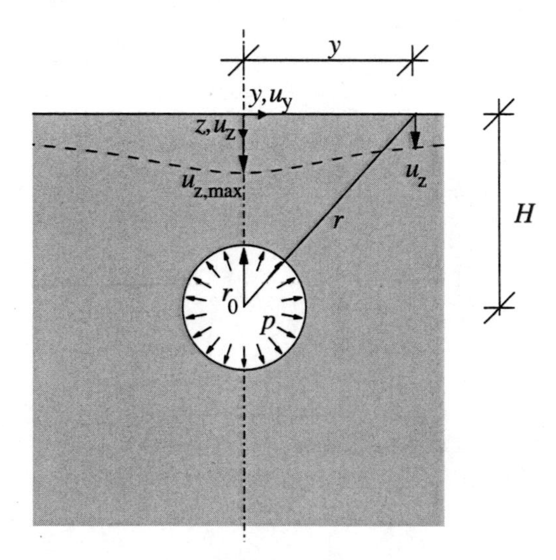

Fig. 2.19: Method of LAMÉ

in a first step. Then the surface is considered by using virtual image technique combined with corrective surface tractions. Virtual image technique means, that a negative mirror image of the sink — i.e. a source — is added (cf. Fig. 2.20). The resulting strains due to the source are added to those in step one. This procedure causes the shear stresses along the surface to vanish. In the third step the remaining normal stresses at the surface are evaluated and subsequently removed by adding stresses in the opposite direction. The resulting strains are again added to those obtained in the steps 1 and 2. The displacement field then reads:

$$
\begin{aligned}
u_y &= -\varepsilon_{r0} r_0 \frac{r_0}{H} \left[y' \left(\frac{1}{r_1'^2} + \frac{1}{r_2'^2} - 4y'z'z_2' \frac{1}{r_2'^4} \right) \right] , \\
u_z &= -\varepsilon_{r0} r_0 \frac{r_0}{H} \left[\frac{z_1'}{r_1'^2} - \frac{z_2'}{r_2'^3} + \frac{2z'(y'^2 - z_2')}{r_2'^4} \right] ,
\end{aligned}
$$

where $z_1 = (z - h)$, $z_2 = (z + h)$, and r_1 and r_2 are the distances to the sink and its image, respectively (see 2.20). The prime ($'$) denotes that the magnitudes are scaled by the tunnel depth H. The radial contraction ε_{r0} may be obtained from $\varepsilon_{r0} = u_0 / r_0$, with u_0 being the radial displacement at the contour line and r_0 being the initial tunnel radius.

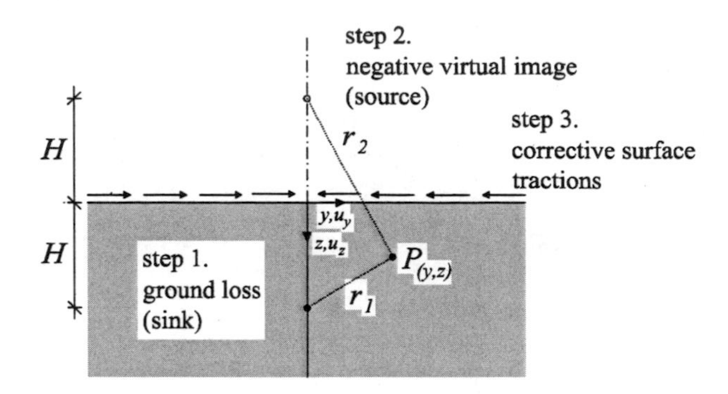

Fig. 2.20: Near surface ground loss using virtual image technique according to [31]

At the soil surface ($z = 0$) the displacements read

$$u_y(z=0) = 2\varepsilon_{r0} r_0 \frac{r_0}{H} \frac{y'}{1+y'^2} \quad ,$$
$$u_z(z=0) = 2\varepsilon_{r0} r_0 \frac{r_0}{H} \frac{1}{1+y'^2} \quad .$$

These equations are very similar to the solution of LAMÉ. But due to the virtual image technique the displacements at the surface are twice those obtained with LAMÉ'S solution.

Additionally ovality and volumetric behaviour can be taken into account. Ovality δ is defined as ratio of the biggest difference of displacements along the circumference $\max(\Delta u) = \max(u(r_0, \theta)) - \min(u(r_0, \theta))$ to the radius r_0 of the lining. The volumetric behaviour is considered by means of the parameter α, which can be chosen in the range from 1.0 to 2.0 [90]. This leads to the following general expressions regarding surface settlements [31]

$$u_y = -2\varepsilon_{r0} r_0 \left(\frac{r_0}{H}\right)^{2\alpha-1} \frac{y'}{(1+y'^2)^\alpha} \left(1 + \frac{\delta}{\varepsilon_{r0}} \frac{1-y'^2}{1+y'^2}\right) \quad ,$$
$$u_z = 2\varepsilon_{r0} r_0 \left(\frac{r_0}{H}\right)^{2\alpha-1} \frac{1}{(1+y'^2)^\alpha} \left(1 + \frac{\delta}{\varepsilon_{r0}} \frac{1-y'^2}{1+y'^2}\right) \quad .$$

Verruijt and Booker [101] proposed a generalisation of SAGASETA's method for compressible soil (arbitrary values of Poisson's ratio μ), that includes the effect of ovality. This solution reads:

$$u_z(y) = 4\varepsilon_{r0} r_0^2 (1-\mu) \frac{H}{y^2+H^2} - 2\delta r_0^2 \frac{H(y^2-H^2)}{(y^2+H^2)^2} \tag{2.3}$$

2.4.1.3 Method of PECK

Compared with the method of LAMÉ, which is not realistic, as measurements show (Atkinson and Potts [8]), the GAUSS-curve proposed by Peck [80] and Schmidt [91] is preferable:

$$u_z(y) = u_{z,\max} \cdot e^{\frac{-y^2}{2i^2}} \quad . \tag{2.4}$$

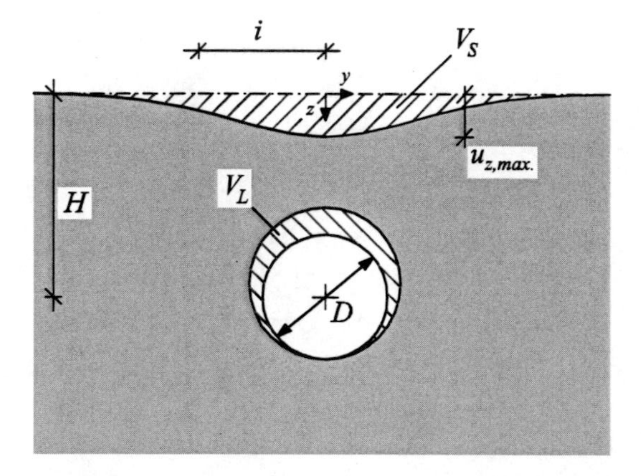

Fig. 2.21: Volume loss V_L and volume of settlement trough V_S.

The volume of the settlement trough per running meter results from the GAUSS-curve to

$$V_S = \int_{-\infty}^{\infty} u_z \mathrm{dy} = \sqrt{2\pi} \cdot i \cdot u_{z,\max} \quad . \tag{2.5}$$

The settlement trough is caused by the fact that a larger amount of soil is excavated than corresponds to the volume of the tunnel. This difference between volume of excavated soil and tunnel volume (defined by the tunnel's outer diameter) is called volume loss or ground loss V_L. In undrained conditions, the volume of the settlement trough V_S equals the volume loss V_L. In general, the volume of the settlement trough is given as ratio of the volume of the settlement trough to the tunnel volume (defined by the tunnel's outer diameter) expressed as a percentage.

If volume of settlement trough V_S and trough width i are known, the maximum settlement $u_{z,\max}$ can be estimated using eq. (2.5) and the settlement trough can be determined. The assessment of these two parameters is crucial and discussed below in detail. Note that most analytical approaches assume undrained condition, which implies the equality of volume of the settlement trough V_S and the volume loss V_L.

Several approaches are mentioned in literature to estimate **volume loss**. E.g., Clough and Schmidt [22] calculate this volume by integrating the theoretical displacements around the tunnel contour, which result from removing the original supporting stresses along the tunnel contour. Assuming $K = 1$ and $\mu \approx 0.5$ (no volume change) the relative volume loss in elasto-plastic material then reads

$$V_L = 2q_u \frac{1+\mu}{E_u} \exp\left(\frac{\sigma_{v0} - p_i}{q_u} - 1\right) \quad , \tag{2.6}$$

with E_u being the elastic modulus of undrained soil and q_u being the unconfined compressive strength. The term $\frac{(\sigma_{v0}-p_i)}{q_u}$ denotes the so called stability ratio N and was proposed by Broms and Bennermark [18]. Model tests [64] and FE calculations showed, that eq. (2.6) is appropriate for deep tunnels, where axisymmetric behaviour can be assumed, but it may not be used for shallow tunnels, where the ground loss cannot be simply related to stability ratio alone.

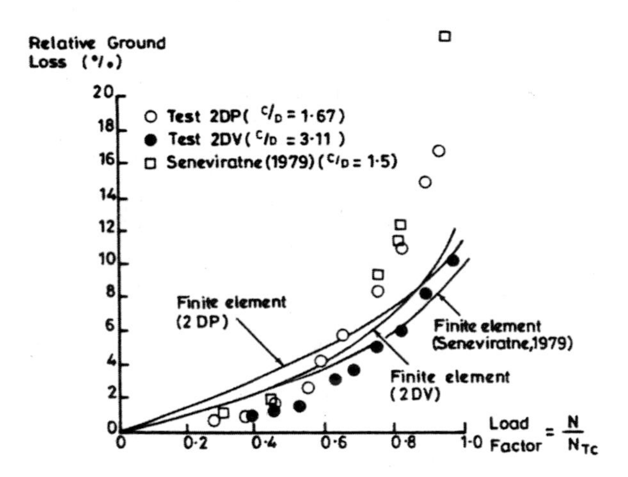

Fig. 2.22: Predicted and observed variation of relative ground loss with load factor according to Mair et al. [64]

To overcome this shortcoming, Mair et al. [64] suggested to introduce the concept of load factor $LF = N/N_{TC}$, with N_{TC} being the stability ratio at collapse.[4] The volume loss can now be obtained from a plot of ground loss

[4]Upper and lower bound values of N_{TC} for plane section tunnels have been derived from plasticity theory by Davis et al. [25].

vs. load factor (e.g. Fig. 2.22) or from a finite element calculation. Gunn [34] pointed out, that there are not sufficient experimental data to support any relationship for volume loss versus load factor for soils apart from kaolin. Hence, this relationship should be handled with care for all other soils.

Macklin [62] derived the following empirical relationship between the volume loss V_L and the stability ratio N from regression analysis of numerous field surveys and lab tests with centrifuge:

$$V_L/A \approx 0.23\ \mathrm{e}^{4.4N/N_{TC}} \quad . \tag{2.7}$$

Beside the mentioned approaches, volume loss is often estimated on the basis of experience with similar tunnelling techniques in similar soils. Mair and Taylor [65] presented the estimated values for V_L given in Tab. (2.4.1.3).

unsupported excavation face in stiff clay:	1-2 %
supported excavation face (slurry or earth mash), sand:	0.5 %
supported excavation face (slurry or earth mash), soft clay:	1-2 %
conventional excavation with sprayed concrete in London clay:	0.5-1.5%

Tab. 2.2: Volume loss of various soil types according to Mair and Taylor [65]

The **trough width** parameter i (standard deviation) equals the y-coordinate at the inflection point of the GAUSS-curve. While Peck [80] and Schmidt [91] proposed i to be dependent on the soil type (Fig. 2.23[5]), Clough and Schmidt [22] give an equation for general application:

$$\frac{2i}{D} = \left(\frac{H}{D}\right)^{0.8} \quad . \tag{2.8}$$

With D being the perimeter of the tunnel and H being the depth of the tunnel axis.

O'Reilly and New [78] suggested that the value of i could be related to the depth of the tunnel axis and a parameter K as

$$i = KH \quad . \tag{2.9}$$

They suggested that K varies from 0.4 to 0.7 for stiff to soft clays and 0.2 to 0.3 for coarse grained materials, as has been generally confirmed by Rankin [83] and Burland et al. [19].

[5] The diagram is based on 17 case studies in the U.S. from 1915 to 1965.

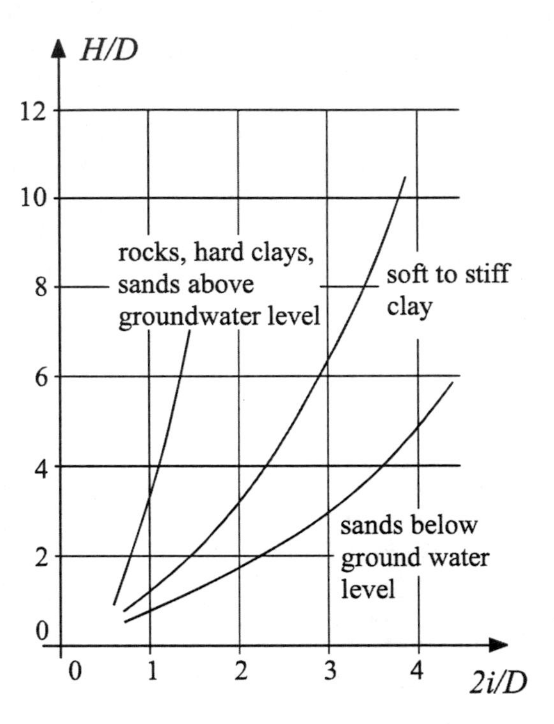

Fig. 2.23: Estimation of i after PECK

The estimations represented here refer to the so-called green field. If the surface is covered by a stiff building, the settlements are smaller [3].

2.4.2 Subsurface settlements

The need to predict subsurface settlements, in particular when considering damage to adjacent structures, has lead to extrapolation of results from surface settlement troughs to the subsurface by replacing eq. (2.9) with

$$i = K(H - z) \tag{2.10}$$

Mair et al. [63] have shown, that the approximations presented for surface settlements can't be used for the prediction of subsurface settlements: the subsurface settlement troughs were considerably wider than predicted by this method. The following relationship for i was proposed by Mair et al. [63]:

$$i/H = 0.175 + 0.325(1 - z/H) \tag{2.11}$$

Grant and Taylor [32] presented data from a number of centrifuge tests. The trough width parameter i shows good agreement with eq. (2.11) apart from a zone in the vicinity of the tunnel where the test data show narrower subsurface troughs, and close to the surface where wider troughs were measured. As a result they define three subsurface areas, where the soil displacement vectors point toward three different focus points (cf. Fig. 2.24).

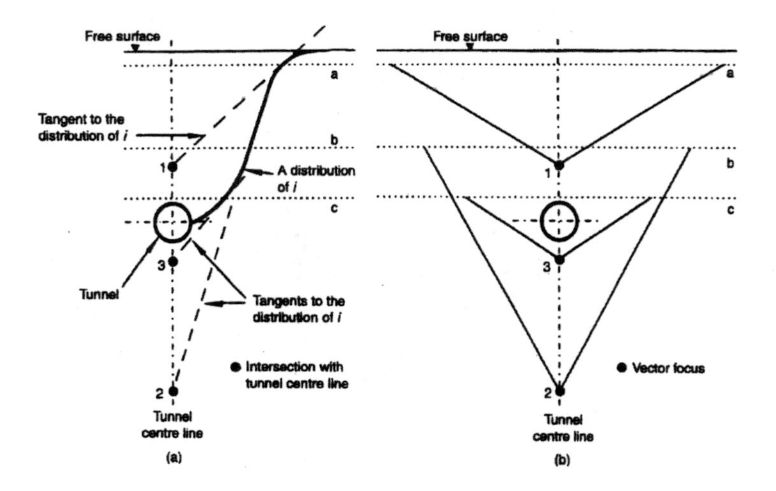

Fig. 2.24: Distribution of i and focus of vectors of soil movements for subsurface settlement troughs with depth [32].

2.4.3 Method of Kolymbas

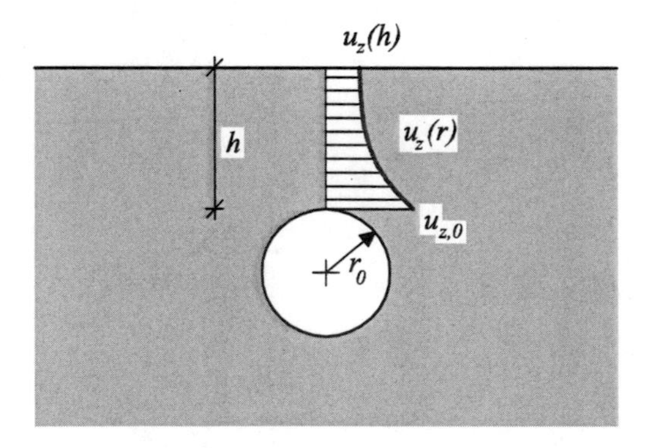

Fig. 2.25: Distribution of the vertical displacement u_z above the crown, according to Kolymbas [51]

Kolymbas [51] proposes the following equation to estimate the displacement distribution above the crown:

$$u_z = u_{z0} \left(\frac{r_0}{r} \right)^d \tag{2.12}$$

with u_{z0} being the vertical displacement at the crown. u_{z0} may be either estimated with the gap method (see section 2.5.2.1, p. 29) or be calculated from

$$u_{z0} = \varepsilon_{\vartheta 0} r_0 = (\varepsilon_{v0} - \varepsilon_{r0}) r_0 \quad . \tag{2.13}$$

ε_{r0} equals the radial strain at the crown and $\varepsilon_{\vartheta 0}$ the tangential strain at the crown. The parameter d is obtained as

$$d = -\frac{\varepsilon_{r0}}{\varepsilon_{\vartheta 0}} \quad . \tag{2.14}$$

The volumetric strain at the crown equals $\varepsilon_v = \varepsilon_{\vartheta 0} + \varepsilon_{r0} = \varepsilon_{r0} \tan \psi$. Thus, radial strain can be calculated using the angle of dilatancy ψ:

$$\varepsilon_{r0} = \frac{\varepsilon_{\vartheta 0}}{\tan \psi - 1} \quad . \tag{2.15}$$

By combining eq. (2.14) and eq. (2.15) the parameter d may be written as

$$d = \frac{1}{1 - \tan \psi} \quad . \tag{2.16}$$

Fig. (2.26) shows the predicted and observed vertical displacements of centrifuge tests conducted by König [48] (see section 2.3.4). The analytical solution obtained with an estimated angle of dilatancy $\psi = 10^\circ$ is in good agreement with the observations.

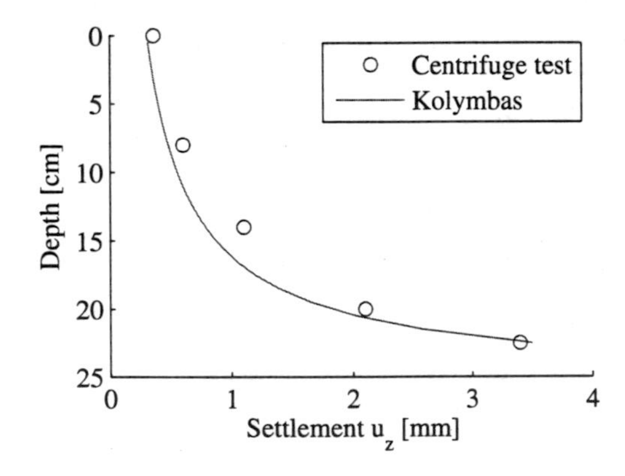

Fig. 2.26: Observed and predicted vertical displacements above centerline from centrifuge tests with non-cohesive soil [48].

2.4.4 Surface settlements in longitudinal direction

Having accepted the GAUSS-curve for the transverse settlement trough, it seems logical to assign a cumulative probability function to the settlement trough on the tunnel centre line.

Attewell and Woodmann [10] start from a point source of loss and assume the GAUSS-curve to be a good assumption to describe transverse settlements. It is further assumed that the soil undergoes no volumetric strain as settlements develop and that ground loss settlements consequent to increments of tunnel advance are additive. In this case the following equations can be written

$$u_z = \frac{V_L}{(2\pi i)^{\frac{1}{2}}} \exp\frac{-y^2}{2i^2}\left[G\left(\frac{x-x_i}{i}\right) - G\left(\frac{x-x_f}{i}\right)\right] \quad , \tag{2.17}$$

$$u_y = \frac{-n}{H-z} y u_z \quad , \tag{2.18}$$

$$u_x = \frac{nV_L}{2\pi(H-z)} \exp\frac{-y^2}{2i^2}\left[\exp\frac{-(x-x_i)^2}{2i^2} - \exp\frac{-(x-x_f)^2}{2i^2}\right] \tag{2.19}$$

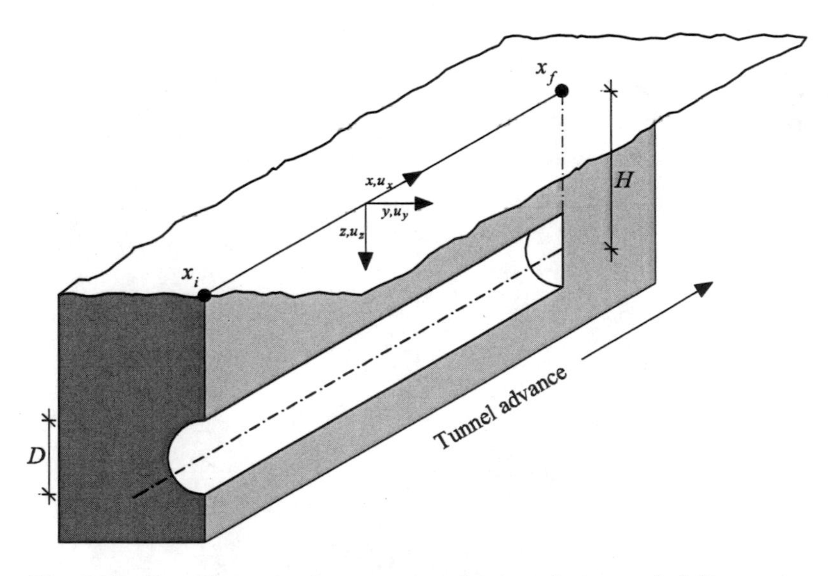

Fig. 2.27: Coordinates and parameters for cumulative probability curve

where V_L is the volume of the transverse settlement trough per unit distance of tunnel advance, i is the distance from the centre line of the tunnel to the point of inflection, n is an empirical parameter defined via:

$$\frac{i}{D} = \frac{1}{2}\left(\frac{H-z}{D}\right)^n \quad , \tag{2.20}$$

H is the depth of soil cover to the tunnel axis, x_i is the initial or tunnel start point, x_f is the position of the face (see Fig. 2.27) and

$$G(\alpha) = \frac{1}{(2\pi)^{\frac{1}{2}}} \int_{-\infty}^{\alpha} \exp \frac{-\beta^2}{2} d\beta. \tag{2.21}$$

G can be found from standard probability tables. In particular $G(0) = 1/2$ and $G(\infty) = 1$.

As Nomoto et al. [75] point out, this assumption is only valid for tunnels constructed with open face-technique, whereas field observations of settlements above tunnel constructed with controlled face pressure (earth pressure shield or slurry shield) show, that the majority of the settlement is associated with the tail void. Therefore, Nomoto proposes the following equation:

$$u_z(y) = ay + by^2 + cy^3 \quad , \tag{2.22}$$

where $u_z(y)$ equals the settlement amount at y and a, b and c are coefficients, which are derived from centrifuge tests. This equation should be handled with care, because one obtains infinite heave/settlement for $y \to \infty$.

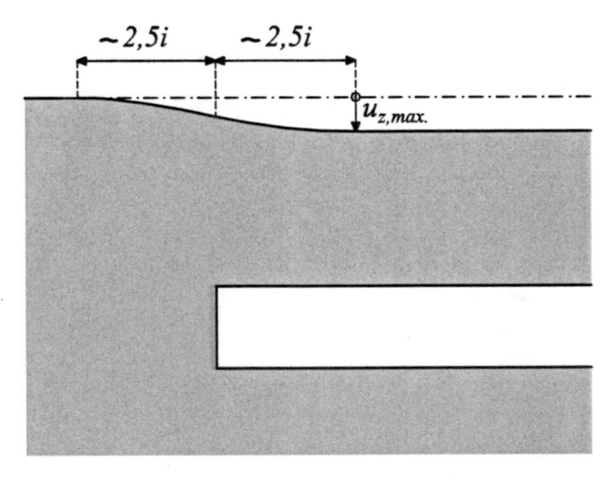

Fig. 2.28: Approximate distribution of the surface settlements in tunnel longitudinal direction.

2.4.5 Summary

The large variety of methods indicates that an analytical method satisfying all requirements has not been found yet. The volume loss method is the most widely used and accepted method to estimate settlement troughs. Although the basic idea of the method, that the volume loss V_L equals the volume of the settlement trough V_S, restrains the method from the use with dilatant soils, the method is applied for the latter as well.

The methods based on continuum mechanics, i.e. method based on LAMÉ's solution and the method of SAGASETA, give usually too wide settlement troughs [90]. SAGASETA [90] presented an approach, that overcomes this shortcoming, by introducing a parameter α, that takes volumetric soil behaviour into account. However, this parameter lacks physical meaning and has to be estimated by trial and error.

Finally the method of Kolymbas is the only method known to the author, that takes volumetric behaviour into account using a physical meaningful parameter: the angle of dilatancy ψ. But the disadvantage is, that one most specify $u_{z,0}$.

A combination of the volume loss method and the method of Kolymbas which reaps the benefits of both is presented in chapter 7.

2.5 Numerical Methods

2.5.1 Introduction

In order to obtain analytical solutions, simplifications and assumptions must be drawn. For example, all analytical solutions are restricted to green field situations and the soil behaviour is usually assumed as isotropic linear elastic and/or ideal plastic. Ground water, tunnel lining and interaction with other subsurface structures (i.e. adjacent tunnels or foundations) are usually neglected.

In contrast, numerical modelling provides the possibility to accommodate the different elements of the interaction problem in one analysis. This section gives an overview of how numerical methods have been applied to predict tunnel induced subsidence. The issue is limited to the Finite Element Method and it will focus on green field analyses.

2.5.2 2D analysis

Tunnel excavation is a 3D problem, but due to the high computational costs of 3D numerical analysis, tunnel excavation is still often modelled in 2D. Most 2D calculations use the idea of volume loss to model tunnelling. They differ in the way volume loss is estimated and the way it is applied.

Shrinkage of lining

Bakker and van Schelt [12] use a two-dimensional approach in which beam elements (to model the lining) are connected to the mesh of continuum elements, which model soil. To develop a specified amount of ground loss (which is chosen arbitrarily to 1%) the lining elements are subjected to a specified amount of circumferential shrinkage. Tamagnini [95] extends this concept by varying the amount of shrinkage of the lining in circumferential direction to simulate the ovalization of the lining.

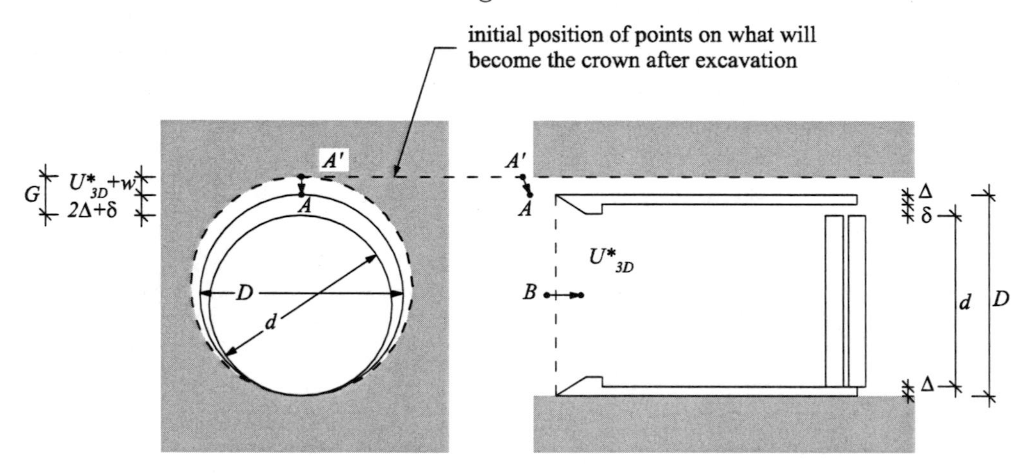

Fig. 2.29: Definition of gap according to Lee and Rowe [60]

The volume loss control method

Addenbrooke and Potts [2] use a stress release procedure, which is related to the the convergence-confinement method introduced by Panet and Guenot [79]. The initial stresses acting on the circumference of the tunnel are incrementally reduced. After each increment the volume of the settlement trough at the surface is calculated. When the prescribed value of the volume of the settlements trough is achieved, the beam elements representing the tunnel lining are placed. If the analysis only focus on the the ground displacement (and no results of the lining stresses and moments are required) it can be terminated after the required volume loss has been reached. This method has been merely used to investigate the behaviour of buildings due to tunnel induced subsidence.

2.5.2.1 The gap method

Rowe and Lo [88] proposed the gap method. The gap G predescribes the final tunnel lining position and size which is smaller than the initial size of the excavation boundary (see Fig. 2.29). Soil movement into the gap is allowed until the soil closes the gap between tunnel and initial excavation boundary position. The gap parameter is the difference between the diameters of the initially excavated boundary and the final tunnel diameter.

The total gap parameter G is composed as follows [61]:

$$G = U_{3D}^{\star} + \omega + 2\Delta + \delta \tag{2.23}$$

with:
- $U_{3D}^{\star}$ ... over-excavation due to 3D movements ahead of the tunnel face,
- ω ... quantity related to workmanship,
- Δ ... thickness of tailskin,
- δ ... clearance required for erection of the lining.

The component $U_{3D}^{\star}$ corresponds to the face loss. Lee [59] proposed, that face loss can be calculated with a plane strain finite element analysis at a longitudinal cross section using the following equation:

$$U_{3D}^{\star} = \frac{k_1}{2} u_{x,\max}^{2\text{D}} \quad ,$$

where k_1 is a factor taking into account the doming effect across the tunnel face (Fig. 2.30) and $u_{x,\max}^{2\text{D}}$ is the maximal intrusion at the tunnel face. k_1 can be calculated as the ratio of the volume of nonuniformal axial intrusion across the tunnel face over the volume assuming uniformal axial intrusion (Fig. 2.30). The parameter k_1 has been examined by Lee [59] using 3D elasto-plastic FE analyses on various combinations of soil parameters and tunnel geometries. Results from these analyses indicate that k_1 is generally in the range of $0.7 - 0.9$.

If the machine is pitched upward or downward, additional material will be excavated. The volume of ground loss over the shield owing to deviations from design grade can be estimated as [23]:

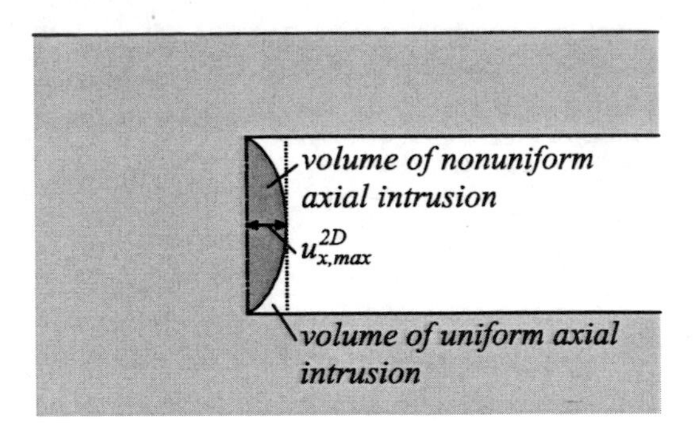

Fig. 2.30: Approximate method to determine face loss $U^{\star}_{3D}$.

$$V_p = \frac{2\pi r_0 L}{2}\lambda \quad , \tag{2.24}$$

with L being the shield length and λ being the angle of excess pitch. From V_p the quantity related to workmanship may be calculated using $\omega = \frac{V_p}{2\pi r_0}$.

$2\Delta + \delta$ is denoted as the *physical gap*. It is determined from geometry once a machine and lining system is chosen.

The gap method is chaotic and many assumptions have to be made to obtain a result. Especially the estimation of the over-excavation $U^{\star}_{3D}$ due to 3D movements ahead of the tunnel face is complex and a separate FE analysis is necessary. Nevertheless, it is the only method known to the author besides 3D FE analyses, that yields an estimation of volume loss V_L.

2.5.2.2 K_0 reduction method

Addenbrooke and Potts [2] found from elasto-plastic axisymmetric analyses of a tunnel heading a reduction in radial stresses while the hoop stresses around the tunnel boundary increased. Hence, he proposed to represent this stress change previous to placement of the tunnel lining by a zone of reduced K_0 at both sides of the tunnel. Analyses which adopted such a zone showed improved settlement profiles [81].

2.5.2.3 Small strain stiffness

Gunn [34] compared analyses using a simple small strain stiffness model with analyses which do not model small strain stiffness. The analyses with small

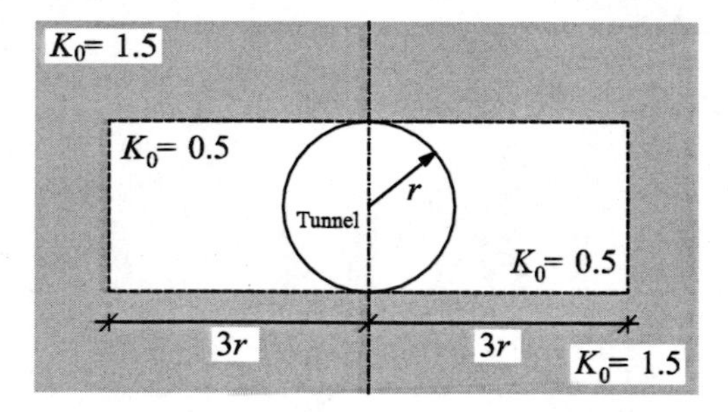

Fig. 2.31: Zone of reduced K_0 [81]

strain stiffness model were much closer to the empirical data on settlement than the ones without that model, although the settlement profiles were still too wide.

2.5.3 3D analysis

Several FE-models have been used to simulate shield tunnelling in 3D. Subsequently some of them are listed and a more or less arbitrary classification in four groups has been undertaken. A brief summary of exemplary models of each group is given and finally the four concepts dealing with 3D calculations are summarised and compared in a table.

2.5.3.1 The gap method

Lee and Rowe [60] extended the model of the gap parameter proposed by Rowe and Kack [87] (see section 2.5.2.1) to a 3D FE-model. In this model the excavation process is simplified in a way, that estimated ground losses are applied along the tunnel axis. That is, the radial convergence over the shield corresponds to the losses over shield (i.e. face loss $U^{\star}_{3D}$ plus losses due to workmanship ω), the convergence behind the shield equals the total gap parameter G (see Fig. 2.32), that includes the physical gap as well.

It is important to mention that this method — despite its 3D nature — does neither include incremental excavation nor an advancement of the shield. As a result, no stress history is taken into account.

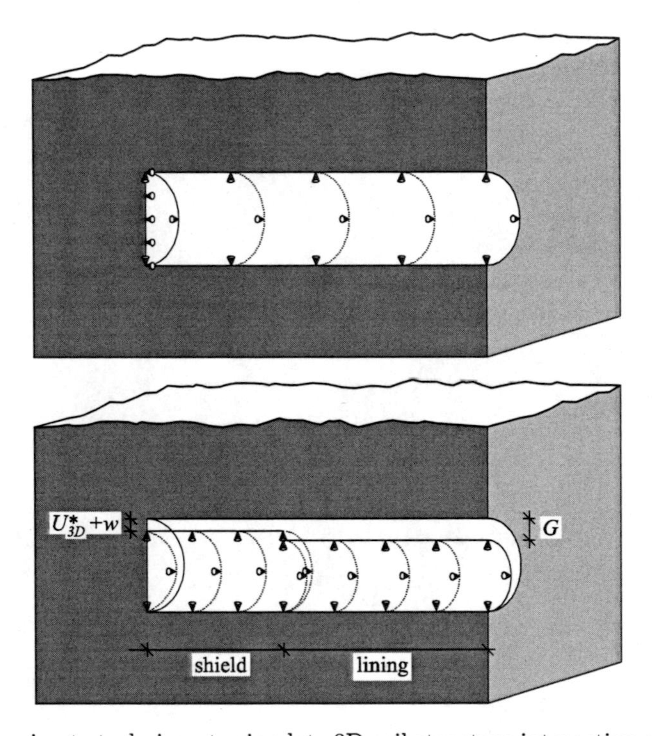

Fig. 2.32: Approximate technique to simulate 3D soil-structure interaction according to Lee and Rowe [60]. Over the shield face loss $U^{\star}_{3D}$ plus losses due to workmanship ω are applied, behind the shield the total gap parameter $G = U^{\star}_{3D} + \omega + 2\Delta + \delta$ is applied.

2.5.3.2 Step-by-step excavation and step-by-step advancement of the shield

Since the FE-method is not able to model continuous excavation, a stepwise excavation and advancement of the shield procedure is usually applied [1, 11, 17, 66]. As a representative, the FE model developed by Mansour [66, see Fig. 2.33] is considered. Beside the stepwise excavation and installation of the lining, this model was the first to take into account the behaviour of the grout.

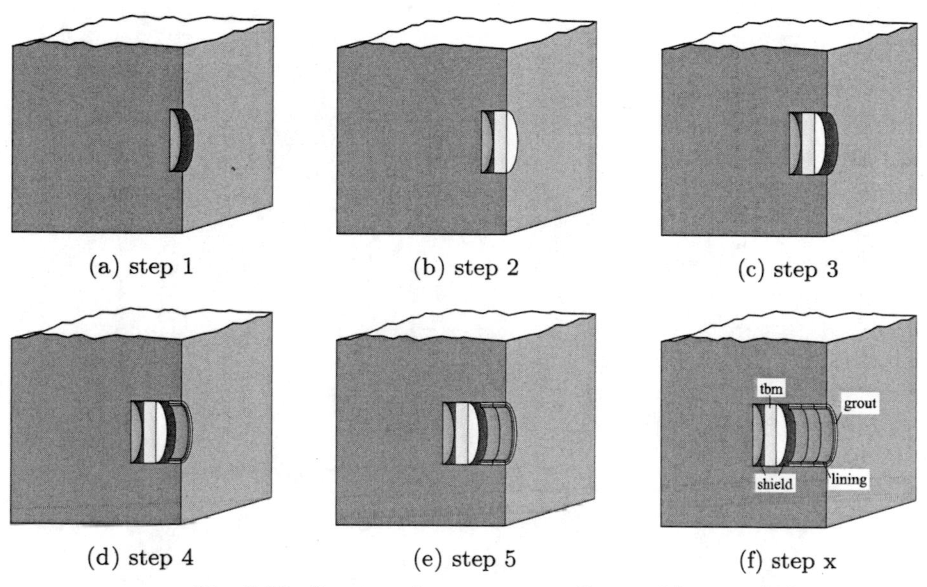

(a) step 1 (b) step 2 (c) step 3

(d) step 4 (e) step 5 (f) step x

Fig. 2.33: Construction steps according to Mansour [66]

The tunnelling process is modelled in the following steps:

Step 0: Inital conditions are applied.

Step 1: Soil elements in the first slice are excavated and elements representing the front part of the shield are activated. The slurry pressure is applied to the tunnel face.

Step 2: Soil elements in the second slice are excavated. The advancement of the machine by one step is represented by activating[6] both the shield

[6] "Activating" means that the stiffness matrix of the activated element, which has been neglected in previous calculation steps, is taken into account for assembling the global stiffness matrix.

elements in the second slice and TBM elements in the first slice. The slurry pressure is applied to the new face.

Step 3: As the previous steps. Besides, the shield tail elements in the third slice are activated.

Step 4: This step includes birth of lining elements and grout elements where they are activated in the first slice. The grouted material in the gap just behind the shield tail is considered in a liquid state and so it has a very low stiffness parameter. Grout pressure is applied.

Step 5: and so forth ...

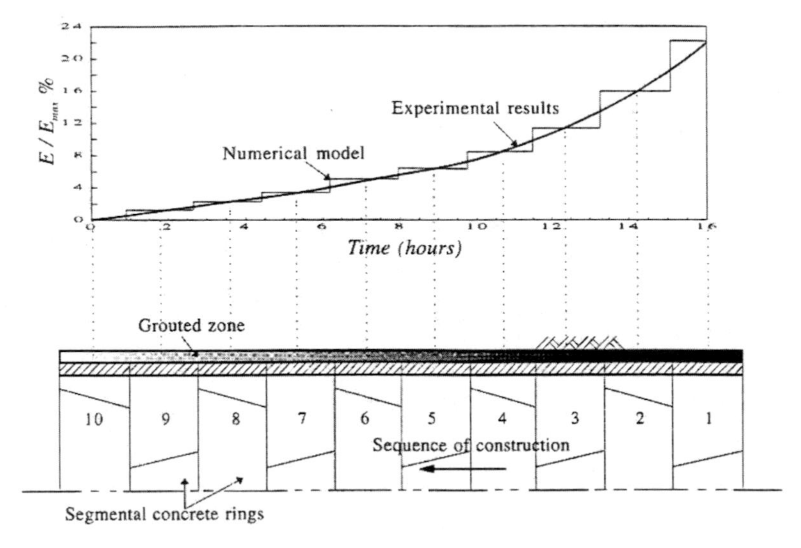

Fig. 2.34: Estimation of relation of elasticity modulus to maximum elasticity modulus for grout material according to advance rate [66]

Soil was modelled using either elasticity or elasto-plasticity with Mohr-Coulomb yield criterion, while lining was modelled assuming elastic behaviour.

As mentioned above, this model was the first one to comprise hardening of the grout. As experimental results show, fresh grout behaves like a liquid, while hardened grout behaves like concrete. To model this time-dependent behaviour, grout is simulated by combining distributed loads (to simulate the pressure when the grout is liquid) and elastic gap elements. While the distributed load gradually vanishes, the stiffness of the grout increases. In the model proposed by Mansour [66] the distributed load, which is a uniform load

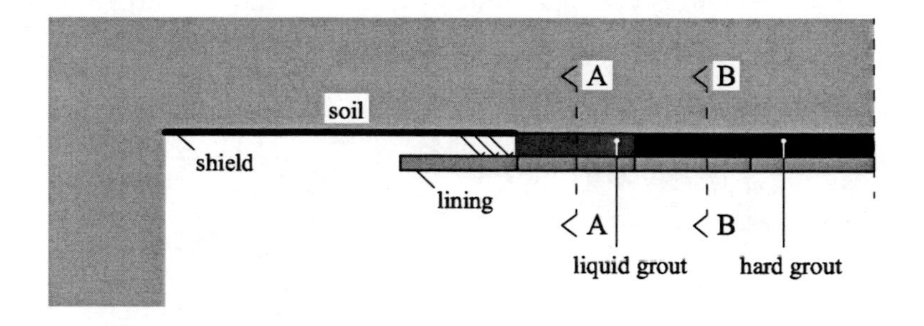

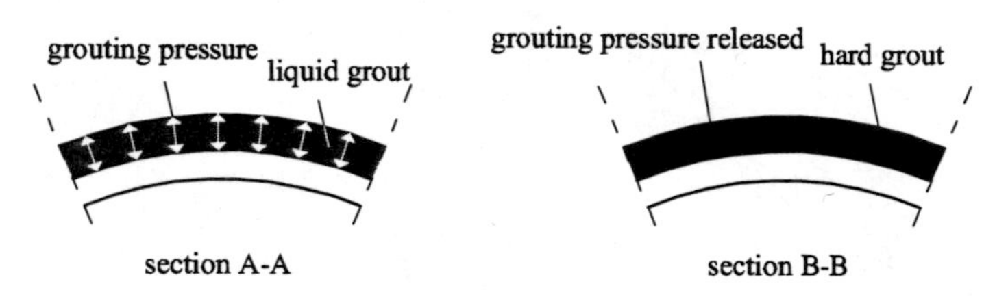

Fig. 2.35: Simulation of grout pressure [66]

on both sides of the gap element, vanishes beyond the first segment behind the shield. Meanwhile the stiffness of the grout elements is increased according to Fig. (2.34), which is obtained by combining a typical time-strength relationship of grout with the advance rate of the shield. The time-strength relationship of grout is obtained from experimental data.

To minimize computational effort, the shield is simulated using equivalent stiffness E_{equiv}. This equivalent stiffness can be estimated as follows:

1. Create a structural system of the machine, with actual geometry and stiffness,

2. analyse the structure under the action of a uniform load P by applying P to the shield to obtain the resulting deformation d_s,

3. replace the structure by square solid elements,

4. vary the stiffness parameters of the solid elements until the model undergoes the same deformation as the actual structure.

Using the described model, Mansour [66] investigated the influence of grout pressure on the soil behaviour by carrying out a parameter study. For various values of grout pressures the final vertical displacements at crown and surface have been plotted against the ratio of grout pressure to overburden pressure (Fig. 2.36).

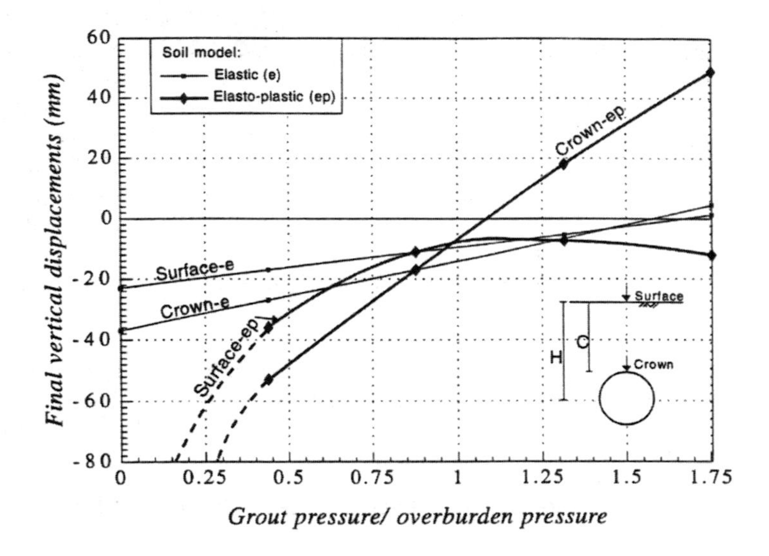

Fig. 2.36: Elastic and elasto-plastic results of final vertical displacements at crown and surface levels vs. grout pressure ratio [66].

While the crown deformation shows the expected behaviour, i.e. decreasing of the vertical displacements with increasing grouting pressure, the surface settlements reduce for very high grout pressures assuming elasto-plastic soil behaviour. MANSOUR explains this effect by a plastified zone, that develops at the crown level (cf. Fig. 2.37) and which allows the overburden to move down.

2.5.4 Excavation Elements

The previously mentioned procedure of stepwise advancement covers the effects of face pressure, lining and grouting, but it is not able to cover the effects of friction between shield and ground nor to consider the forces of the hydraulic jacks.

To simulate friction between shield and ground the concept of "Excavating Elements" [6, 7, 53, 54] consideres ground and shield as two bodys, which are connected with joint elements. The advancement of the shield is controlled by the hydraulic jack forces applied at the tail of the shield (see Fig. 2.38). In

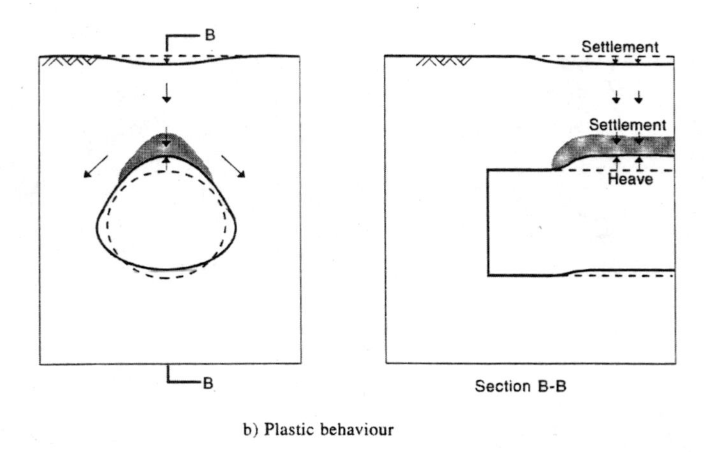

Fig. 2.37: Effect of excessive grouting pressure in FE analysis with elasto-plastic material model: above the crown a plastified zone develops due to grouting. The zone is compressed and the soil above this zone is unaffected by tail gap grouting. [66].

front of the shield solid finite elements of a fixed size are introduced, which aim to represent the disturbed ground in front of the face. Size and material properties of these elements are determined by trial and error in a way that the computed data fit with the monitored data of the construction site, i.e. the advancement of the virtual shield at a given time step fits to the measured advancement of the real shield [54] [7]. Due to the large deformations in front of the shield, remeshing is essential.

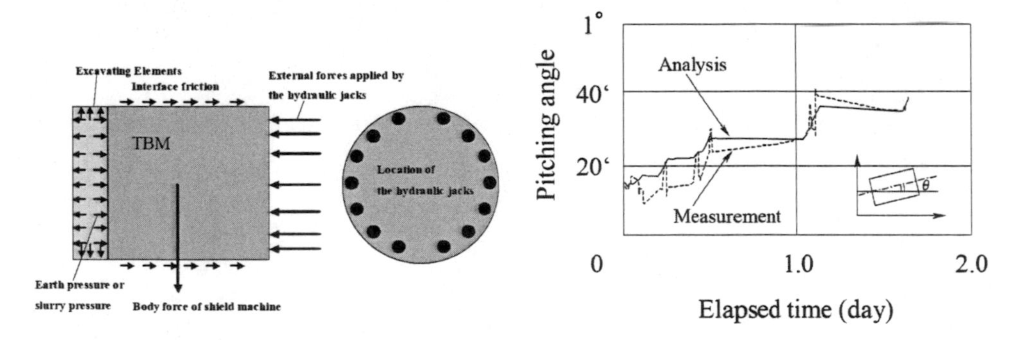

Fig. 2.38: Modelling shield tunnelling with "excavating elements" [53]

Fig. 2.39: Pitching angle of the shield machine [53]

[7] As the advancement is not time dependent in elasto-plastic materials, this method of parameter fitting is confusing. Bearing in mind, that the jacking forces increase from zero to the recorded value during one time step of the FEM calculation, the stiffness has to be fitted to the recorded distance the shield advanced during the time measured afield.

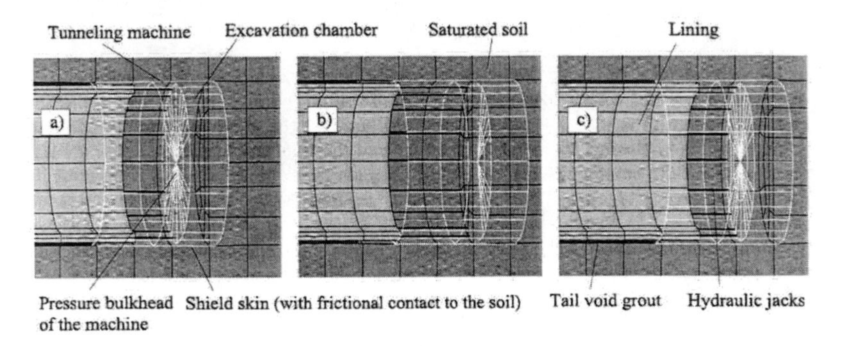

Fig. 2.40: Components and step-by-step simulation procedure of shield tunnelling [47]

The slurry pressure applied during the construction is considered as a static boundary condition upon the excavation elements (see Fig. 2.38). The grouting process is simulated by applying to the mesh adjacent to the tunnel a pressure equivalent to the measured grouting pressure.

The results of Komiya [53] fit very well to the measurements. Especially the pitching angle shows reasonable agreement (see Fig. 2.39). But one has to be aware of the arbitrariness of the *excavating elements*. They are a nice tool to fit results, but their lack of physical meaning will cause problems in determining parameters, not to mention the impossibility of class A predictions.

2.5.5 Step-by-step excavation combined with continuous advancement of the shield

Kasper and Meschke [47] put all the previously mentioned aspects together and created a sophisticated model, which takes into account groundwater, frictional contact, excavation and support of the soil, hydraulic jacks, installation of the tunnel lining and grout injection of the tail void. The Cam-Clay plasticity model is used to describe the soil behaviour, grout is assumed to be a material with time dependent Young's modulus and permeability. Both are modeled as saturated porous materials, using a two-field finite element formulation.

The tunnelling process is simulated in a step-by-step procedure (see Fig. 2.40). In the first step the shield is pushed ahead by a length change of truss elements representing the jacks. Then the excavation is simulated by re-zoning the

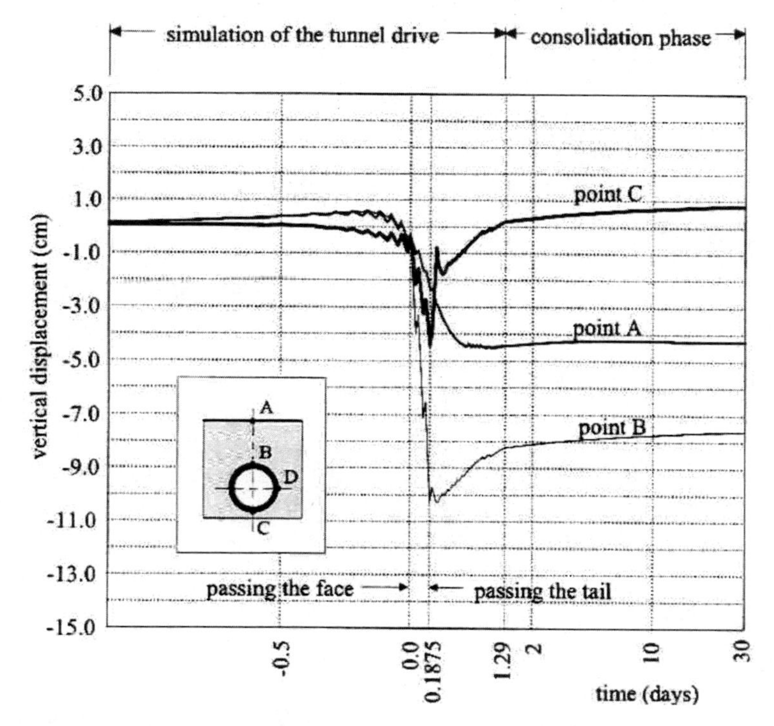

Fig. 2.41: Predicted vertical soil displacements at the ground surface (point A), at the tunnel crown (point B) and at the tunnel invert point (point C) [47]

finite element mesh at the face, i.e. removing elements, relocating the nodes on the prescribed excavation boundary and transferring all variables from the old to the new mesh. At the same time, new lining rings are installed and tail void grouting is modelled by introducing elements representing the grout. The face is supported by an applied pressure, that acts on the face as well as on the shield. A pore water pressure is prescribed directly behind the shield corresponding to the grout injection pressure.[8] The back-up trailer is simulated by applying point loads to the lining.

This model is very complex and takes into account most of the relevant factors that influence the soil displacements. In the author's view, one point should be improved: The vertical movement of the crown along the shield doesn't correspond to field measurements (cf. Fig. 2.5 with Fig. 2.41, point B). While the field measurements depict an outward soil displacement due to

[8]Kasper and Meschke [47] do not clearly mention to which distance from the tail the pore water pressure is applied. Looking at the results of the excess pore pressure they plot, one can assume, that pore pressure is only prescribed to the elements very next to the shield.

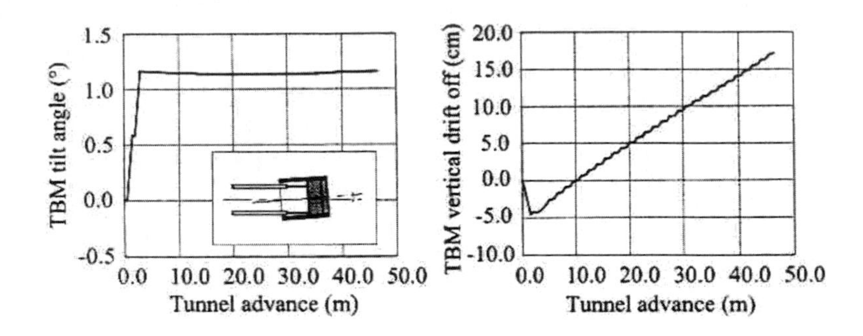

Fig. 2.42: Computed shield movement: a) inclination and b) vertical drift off

the grouting pressure, the soil moves toward the tunnel axis in the analysis. This shortcoming may be overcome by applying grouting pressure along the shield.

2.5.6 Summary

This section presents a number of 2D- and 3D-FE calculations. The following points are highlighted by their authors:

- It has been noted by many authors that the process of tunnel excavation is clearly a 3D problem. However, due to limitations in computational power 2D analysis is widely used. For plane strain analyses there are different approaches to account for stress redistribution ahead of the tunnel face.
- Modelling technique (i.e. boundary conditions) has reached a high-level and has been analysed in detail. However, the influence of constitutive model has not yet been sufficiently investigated.
- Analyses with small strain stiffness model were much closer to the empirical data than the ones without that model.

2.6 Conclusions

The previous subsections present different approaches to estimate tunnel induced ground deformation.The FE method proves to be the most powerful but also most costly method. Summaraising model tests and numerical and analytical analyses, the following requirements may be rised for a FE model to estimate surface settlements:

Construction process should be simulated in detail, as e.g. Komiya [53] or Kasper and Meschke [47] did. On the one hand it was shown, that the results are perceivably influenced by the way the construction process is simulated. On the other hand a detailed simulation of the construction process avoids the estimation of volume loss, which is the critical point in most methods.

Rigid body failure mechanism: As rigid body failure mechanism dominates the deformation patterns, FE analyses should be able to reproduce them. This may be achieved by remeshing when localization occurs or by FE codes that model strong discontinuities.

3D nature of the tunnelling process must be accounted for.

Volumetric soil behaviour at cyclic loading must be considered by the usage of appropriate constitutive models.

Small strain stiffness: The constitutive model should reproduce small strain stiffness.

Two-phase medium (solid grains and pore water) should be considered for problems comprising groundwater.

Chapter 3

Numerical modelling

All numerical simulations in this thesis are performed with ABAQUS, version 6.4. This commercial FE code provides an interface to include constitutive models, the so-called UMAT. The task of the UMAT is to supply ABAQUS with the new Cauchy stress tensor $\sigma(t_a + \Delta t)$, updated according to the constitutive model, as well as with the Jacobian matrix $\frac{\partial \sigma_{ij}}{\partial \varepsilon_{kl}}$ and the updated state variables. The precise evaluation of the Jacobian is essential to achieve fast (quadratic) convergence.

The UMAT used in this thesis is an extended version of the one written by Fellin and Ostermann [29], who used a second-order extrapolated Euler method in their UMAT that achieves quadratic convergence. This UMAT was extended by the author by the concept of intergranular strains to take small strain stiffness into account.

Starting with the basics of the Finite Element method and the nonlinear solution technique of ABAQUS, the implementation of the constitutive model is outlined.

Thereafter a short introduction to hypoplasticity and the concept of intergranular strain is given and numerical problems arising from the implementation are discussed.

Finally, the appropriateness of hypoplasticity and Mohr-Coulomb material model for simulation of tunnelling are examined by means of element tests and FE analyses.

3.1 Variational approach to the Finite Element Method

The governing equilibrium equation in continuum mechanics in a body with applied volume forces of $\mathbf{f} = \rho \mathbf{g}$ reads

$$\operatorname{div} \boldsymbol{\sigma} + \mathbf{f} = \mathbf{0} \quad . \tag{3.1}$$

Together with the boundary conditions

$$\mathbf{t} - \boldsymbol{\sigma}\mathbf{n} = \mathbf{0} \quad . \tag{3.2}$$

the problem can be solved. $\boldsymbol{\sigma}$ denotes the symmetric Cauchy-stress, $\mathbf{t}$ is the stress vector acting on the boundary surface A with $\mathbf{n}$ being the outward unit normal vector.

Transformation of eq. (3.1) into its weak representation, which is obtained by multiplying eq. (3.1) and eq. (3.2) with an arbitrary $\delta\mathbf{u}$, integrating over the volume V and the surface A and following addition of the terms, yields

$$\int_V \delta\mathbf{u}^T \operatorname{div}\boldsymbol{\sigma}\,\mathrm{dV} + \int_V \delta\mathbf{u}^T \mathbf{f}\,\mathrm{dV} + \int_A \delta\mathbf{u}^T(\mathbf{t} - \boldsymbol{\sigma}\mathbf{n})\,\mathrm{dA} = 0 \quad , \tag{3.3}$$

assuming an equilibrated initial state. Using the chain rule and the divergence theorem, the first term in eq. (3.3) can be written as

$$\int_V \delta\mathbf{u}^T \operatorname{div}\boldsymbol{\sigma}\,\mathrm{dV} = \int_A \delta\mathbf{u}^T\boldsymbol{\sigma}\mathbf{n}\,\mathrm{dA} - \int_V \operatorname{grad}\delta\mathbf{u} : \boldsymbol{\sigma}\,\mathrm{dV} \quad . \tag{3.4}$$

Substituting the first term of eq. (3.3) by eq. (3.4), the virtual work formulation is achieved

$$\int_V \delta\boldsymbol{\varepsilon} : \boldsymbol{\sigma}\,\mathrm{dV} = \int_A \delta\mathbf{u}\mathbf{t}\,\mathrm{dA} + \int_V \delta\mathbf{u}\mathbf{f}\,\mathrm{dV} \quad . \tag{3.5}$$

3.1.1 Discretisation in space

To solve eq. (3.5), the investigated body is subdivided in an adequate number of finite elements. The elements are assumed to be interconnected at a discrete number of nodal points. A set of functions (so called shape functions) is chosen to approximate the displacement within each finite element. The displacements within the elements are given by

$$\mathbf{u}^e = \mathbf{N}^e\mathbf{q}^e, \tag{3.6}$$

where $\mathbf{q}^e$ represents the displacements at the nodal points and $\mathbf{N}^e$ are the so called shape functions. With the displacements known within the element the strain can be determined using the matrix operator $\mathbf{B}$, which results from multiplying the shape function $\mathbf{N}^e$ by the differential operator $\mathbf{L}$. The matrix operator $\mathbf{B}$ of a 2D linear triangular element, which is frequently used in this work, reads:

$$\mathbf{B}^e = \begin{bmatrix} \frac{\partial N^{(i)}}{\partial x} & 0 & \frac{\partial N^{(j)}}{\partial x} & 0 & \frac{\partial N^{(k)}}{\partial x} & 0 \\ 0 & \frac{\partial N^{(i)}}{\partial y} & 0 & \frac{\partial N^{(j)}}{\partial y} & 0 & \frac{\partial N^{(k)}}{\partial y} \\ \frac{\partial N^{(i)}}{\partial y} & \frac{\partial N^{(i)}}{\partial x} & \frac{\partial N^{(j)}}{\partial y} & \frac{\partial N^{(j)}}{\partial x} & \frac{\partial N^{(k)}}{\partial y} & \frac{\partial N^{(k)}}{\partial x} \end{bmatrix} , \tag{3.7}$$

with i, j, k representing the three nodes of the element. The strain can now be interpolated via

$$\boldsymbol{\varepsilon}^e = \mathbf{B}^e \mathbf{q}^e \quad . \tag{3.8}$$

The vector of nodal displacements of one element is connected to the vector of nodal displacements of the whole system via the incidence matrix $\mathbf{Z}$:

$$\mathbf{q}^e = \mathbf{Z}^e \mathbf{q} \quad . \tag{3.9}$$

Using the index $e = 1, \ldots, m$ to denote the elements, eq. (3.5) can be written as

$$\sum_{e=1}^{m} \int_{V^e} (\mathbf{B}^e \mathbf{Z}^e \delta \mathbf{q})^T \boldsymbol{\sigma} \, \mathrm{dV} = \sum_{e=1}^{m} \int_{A^e} (\mathbf{N}^e \mathbf{Z}^e \delta \mathbf{q})^T \mathbf{t} \, \mathrm{dA} + \sum_{e=1}^{m} \int_{V^e} (\mathbf{N}^e \mathbf{Z}^e \delta \mathbf{q})^T \mathbf{f} \, \mathrm{dV} \tag{3.10}$$

V^e and A^e denote the volume and area of an element. As eq. (3.5) is valid for any vector $\delta\mathbf{u}$, $\delta\mathbf{u}$ may be cancelled and eq. (3.10) reads

$$\underbrace{\sum_{e=1}^{m} \mathbf{Z}^{e^T} \int_{V^e} \mathbf{B}^{e^T} \boldsymbol{\sigma} \, \mathrm{dV}}_{\mathbf{r}_{int}} = \underbrace{\sum_{e=1}^{m} \mathbf{Z}^{e^T} \int_{A^e} \mathbf{N}^{e^T} \mathbf{t} \, \mathrm{dA} + \sum_{e=1}^{m} \mathbf{Z}^{e^T} \int_{V^e} \mathbf{N}^{e^T} \mathbf{f} \, \mathrm{dV}}_{\mathbf{r}_{ext}} , \tag{3.11}$$

with $\mathbf{r}_{int}$ being the internal force vector and $\mathbf{r}_{ext}$ denoting the external force vector. Eq. (3.11) is solved using numerical integration [13, 105].

3.1.2 Discretisation in time

Due to the nonlinear constitutive model and the nonlinear geometry imposed by large displacements, an incremental-iterative approach for the solution of boundary value problems must be followed. That is, external forces and/or prescribed displacements are applied in a number of increments. Within each increment equilibrium is reached in an iterative manner. Starting from an equilibrium at time t_a stress history is discretised as follows:

$$\boldsymbol{\sigma}^{t_a+\Delta t} = \boldsymbol{\sigma}^{t_a} + \Delta\boldsymbol{\sigma}^{t_a} \quad , \tag{3.12}$$

For the time $t_a + \Delta t$, eq. (3.11) can be rewritten as

$$\begin{aligned} &\sum_{e=1}^{m} \mathbf{Z}^{e^T} \int_{V^e} \mathbf{B}^{e^T} \boldsymbol{\sigma}^{t_a} \, \mathrm{dV} + \sum_{e=1}^{m} \mathbf{Z}^{e^T} \int_{V^e} \mathbf{B}^{e^T} \Delta\boldsymbol{\sigma}^{t_a} \, \mathrm{dV} = \\ &\sum_{e=1}^{m} \mathbf{Z}^{e^T} \int_{A^e} \mathbf{N}^{e^T} \mathbf{t}^{t_a+\Delta t} \, \mathrm{dA} + \sum_{e=1}^{m} \mathbf{Z}^{e^T} \int_{V^e} \mathbf{N}^{e^T} \mathbf{f}^{t_a+\Delta t} \, \mathrm{dV} \end{aligned} \tag{3.13}$$

Geometrical nonlinearity means, that the volume V^e and the area A^e are unknown at time $t + \Delta t$, while material nonlinearity stands for $\Delta\boldsymbol{\sigma}$ being a nonlinear function of $\Delta\boldsymbol{\varepsilon}$ (and consequently of $\Delta\mathbf{u}$). Eq. (3.13) can be rewritten as

$$\begin{aligned} &\underbrace{\sum_{e=1}^{m} \mathbf{Z}^{e^T} \int_{V^e} \mathbf{B}^{e^T} \Delta\boldsymbol{\sigma}^{t_a} \, \mathrm{dV}}_{\Delta\mathbf{r}_{int}^{t_a}} = \\ &= \underbrace{\sum_{e=1}^{m} \mathbf{Z}^{e^T} \int_{A^e} \mathbf{N}^{e^T} \mathbf{t}^{t_a+\Delta t} \, \mathrm{dA} + \sum_{e=1}^{m} \mathbf{Z}^{e^T} \int_{V^e} \mathbf{N}^{e^T} \mathbf{f}^{t_a+\Delta t} \, \mathrm{dV}}_{\mathbf{r}_{ext}^{t_a+\Delta t}} - \\ &\qquad - \underbrace{\sum_{e=1}^{m} \mathbf{Z}^{e^T} \int_{V^e} \mathbf{B}^{e^T} \boldsymbol{\sigma}^{t_a} \, \mathrm{dV}}_{\mathbf{r}_{int}^{t_a}} \end{aligned} \tag{3.14}$$

To solve eq. (3.14), the nonlinear material model has to be written in incremental form

$$\Delta\boldsymbol{\sigma}^{t_a} = \Delta\boldsymbol{\sigma}^{t_a}(\boldsymbol{\sigma}(t_a), \Delta\boldsymbol{\varepsilon}(\Delta\mathbf{u}), \mathbf{Q}) \quad . \tag{3.15}$$

with Q denoting the additional state variables. Thus we obtain

$$\Delta\boldsymbol{\sigma}^{t_a} = \underbrace{\frac{\partial\Delta\boldsymbol{\sigma}}{\partial\Delta\boldsymbol{\varepsilon}}}_{\mathcal{M}} \cdot \Delta\boldsymbol{\varepsilon} = \mathcal{M}\mathbf{B}^e\Delta\mathbf{q} = \mathcal{M}\mathbf{B}^e\mathbf{Z}^e\Delta\mathbf{q} \quad . \tag{3.16}$$

Eq. (3.14) now reads

$$\sum_{e=1}^{m} \mathbf{Z}^{e^T} \int_{V^e} \mathbf{B}^{e^T} \mathcal{M} \mathbf{B}^e \mathbf{Z}^e \Delta\mathbf{q} \, \mathrm{dV} = \mathbf{r}_{ext}^{t_a+\Delta t} - \mathbf{r}_{int}^{t_a} \quad . \tag{3.17}$$

Using the global tangent stiffness matrix

$$\mathbf{K} = \sum_{e=1}^{m} \mathbf{Z}^{e^T} \int_{Ve} \mathbf{B}^{e^T} \mathcal{M} \mathbf{B}^e \, \mathrm{dV} \; \mathbf{Z}^e \quad , \tag{3.18}$$

eq. (3.17) can be rewritten as

$$\mathbf{K}\Delta\mathbf{q} = \mathbf{r}_{ext}^{t_a+\Delta t} - \mathbf{r}_{int}^{t_a} \quad . \tag{3.19}$$

The kinematic boundary conditions for the displacements, e.g. the fixed nodes, are considered in eq. (3.19) through an appropriate elimination of certain rows and columns of the stiffness matrix $\mathbf{K}$ and of the corresponding rows of the residuum. The system is thereby reduced to the remaining degrees of freedom.

3.1.3 Nonlinear solution technique

The nonlinear equilibrium equations (3.19) are solved using an incremental-iterative solution technique[1], where the total load is applied in increments (see Fig. 3.1). The equations are solved for each increment Δt within several iterations. Starting from the known state at the end of iteration j, the state of iteration $j+1$ is achieved by accomplishing the following steps:

1. Correction of displacement vector:

$$\Delta\mathbf{q}_{j+1} = \mathbf{K}_j^{-1} (\mathbf{r}_{ext}^{t_a+\Delta t} - \mathbf{r}_{int,j}^{t_a+\Delta t}) \quad .$$

2. Updating displacements:

$$\mathbf{q}_{j+1} = \mathbf{q}_j + \Delta\mathbf{q}_{j+1} \quad .$$

3. Incremental strains of iteration $j+1$ follow as:

$$\Delta\boldsymbol{\varepsilon}_{j+1} = \mathbf{B}\Delta\mathbf{q}_{j+1} \quad .$$

[1] ABAQUS generally uses the Newton-Raphson method.

4. Having calculated the strain increments, a subroutine is now able to integrate the state variables $(\boldsymbol{\sigma}, e, \mathbf{S})$ over the time Δt_{j+1} as well as to compute the derivative $\mathcal{M}_{j+1}$ of the stress with respect to strain increments:

$$\boxed{\text{UMAT} \rightarrow} \begin{cases} \mathcal{M}_{j+1} & = \dfrac{\partial \Delta \boldsymbol{\sigma}_{j+1}}{\partial \Delta \boldsymbol{\varepsilon}_{j+1}} \quad , \\ \Delta \boldsymbol{\sigma}_{j+1} & = \mathcal{M}_{j+1} \Delta \boldsymbol{\varepsilon}_{j+1} \quad , \\ \boldsymbol{\sigma}_{j+1} & = \boldsymbol{\sigma}_j + \Delta \boldsymbol{\sigma}_{j+1} \quad . \end{cases}$$

5. Updating internal forces:

$$\mathbf{r}_{int,j+1}^{t_a+\Delta t} = \sum_{e=1}^{m} \mathbf{Z}^{e^T} \int_{V^e} \mathbf{B}_j^{e^T} \boldsymbol{\sigma}_{j+1} \, \mathrm{dV} \quad .$$

6. Consistent global stiffness matrix $\mathbf{K}_{j+1}$ is assembled:

$$\mathbf{K}_{j+1} = \sum_{e=1}^{m} \mathbf{Z}^{e^T} \int_{Ve} \mathbf{B}^{e^T} \mathcal{M}_{j+1} \mathbf{B}^e \, \mathrm{dV} \; \mathbf{Z}^e \quad .$$

7. Now the residual forces can be calculated and the convergence criteria are checked:

$$\begin{aligned} |\mathbf{r}_{ext}^{t_a+\Delta t} - \mathbf{r}_{int,j+1}^{t_a+\Delta t}| & \leq \Delta r_{tol} \\ |\Delta \mathbf{q}_{j+1}| & \leq \Delta q_{tol} \end{aligned}$$

If they are fulfilled, the next increment is started, otherwise another iteration is necessary.

Depending on when and how often the global stiffness matrix $\mathbf{K}$ is updated, the name of the Newton-type solution technique varies [41]. The *initial stiffness matrix method* is the simplest and a very robust method, where $\mathbf{K}$ is only assembled at the beginning of the calculation. In contrast, in the *Newton-Raphson method* $\mathbf{K}$ is updated for every iteration in every increment. This causes an increase of convergence, but a decrease of robustness. In the *modified Newton-Raphson* method $\mathbf{K}$ is updated in every increment, but not in every iteration. Lastly, the quasi-Newton method uses an approximated secant stiffness instead of the tangent stiffness [93].

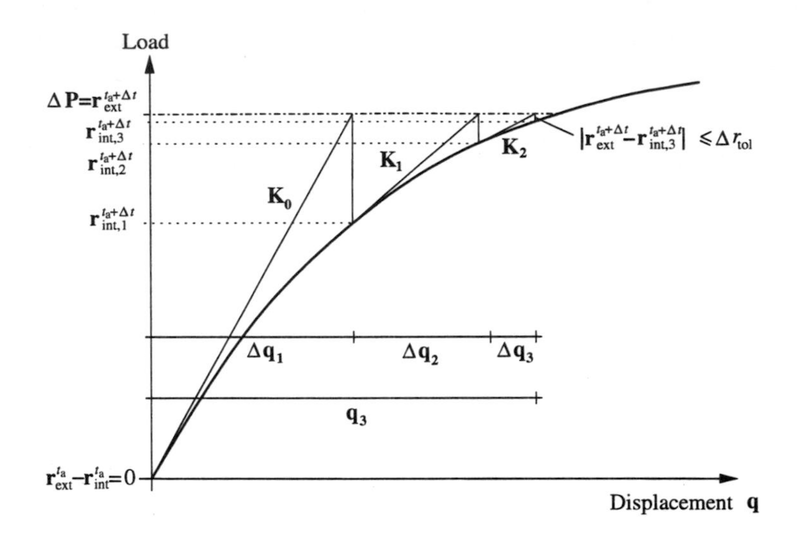

Fig. 3.1: NEWTON-RAPHSON iteration.

3.1.4 Time integration of state variables and determination of the Jacobian matrix

The equilibrium equation, together with the constitutive model lead to a coupled system consisting of an initial boundary value problem and an ordinary differential equation, that can be solved by co-simulation: the equilibrium equations are solved by the FE-program and the constitutive model by an external solver, e.g. UMAT in ABAQUS [28]. In the above listed iteration cycle the task of the UMAT is to provide the new stress tensor $\boldsymbol{\sigma}(t_a + \Delta t) = \mathbf{T}(\Delta t)$ at time $t_a + \Delta t$ as well as its derivative with respect to the strain increment $\Delta\boldsymbol{\varepsilon}$ for the given initial stress tensor $\mathbf{T}(0) = \boldsymbol{\sigma}(t_a)$. Thus the following system of differential equations for $0 \leq t \leq \Delta t$ has to be solved:

$$\frac{\mathrm{d}}{\mathrm{dt}}\mathbf{T} = \mathbf{h}(\mathbf{T}, \mathbf{D}, \mathbf{Q}), \qquad \mathbf{T}(0) = \boldsymbol{\sigma}(t_a) \quad , \tag{3.20}$$

$$\frac{\mathrm{d}}{\mathrm{dt}}\mathbf{Q} = \mathbf{k}(\mathbf{T}, \mathbf{D}, \mathbf{Q}), \qquad \mathbf{Q}(0) = \mathbf{Q}(t_a) \quad , \tag{3.21}$$

where $\mathbf{Q}$ denotes the additional state variables and $\mathbf{D}$ the stretching. The latter is defined as:

$$\mathbf{D} = \frac{1}{2}\left(\operatorname{grad}\dot{\mathbf{u}} + (\operatorname{grad}\dot{\mathbf{u}})^T\right) \quad .$$

The calculation starts from an equilibrium at time t_a knowing the Cauchy stress tensor $\mathbf{T}$ as well as the time increment Δt and an initial guess $\Delta\boldsymbol{\varepsilon}_0$ of the strain increment. To assure quadratic convergence of the Newton-Raphson method, the precise determination of the Jacobian

$$\mathcal{M}_{j+1} = \frac{\partial \Delta \boldsymbol{\sigma}_{j+1}}{\partial \Delta \boldsymbol{\varepsilon}_{j+1}} \tag{3.22}$$

is essential.

The Jacobian of the material model $\mathcal{M}$ is given by

$$\frac{\partial \Delta \boldsymbol{\sigma}}{\partial \Delta \boldsymbol{\varepsilon}} = \frac{\partial \boldsymbol{\sigma}(t + \Delta t)}{\partial \Delta \boldsymbol{\varepsilon}} = \frac{1}{\Delta t} \cdot \frac{\partial \mathbf{T}}{\partial \mathbf{D}}(\Delta t) \quad , \tag{3.23}$$

which depends on the global iterative solution technique and on the time integration scheme of the material model [41]. The Jacobian can roughly be approximated using numerical differentiation at the end of the iteration increment [85] or it can be calculated for every sub-iteration increment and then be integrated over time [28, 41, 70]. In this work the following approximation obtained by numerical differentiation was used [28]:

$$\frac{\mathrm{d}}{\mathrm{dt}} \mathbf{B}_{ij} = \frac{1}{\vartheta} \big(\mathbf{h}(\mathbf{T} + \vartheta \mathbf{B}_{ij}, \mathbf{D} + \vartheta \mathbf{V}_{ij}, \mathbf{Q} + \vartheta \mathbf{G}_{ij}) - \mathbf{h}(\mathbf{T}, \mathbf{D}, \mathbf{Q}) \big) \tag{3.24}$$

$$\frac{\mathrm{d}}{\mathrm{dt}} \mathbf{G}_{ij} = \frac{1}{\vartheta} \big(\mathbf{k}(\mathbf{T} + \vartheta \mathbf{B}_{ij}, \mathbf{D} + \vartheta \mathbf{V}_{ij}, \mathbf{Q} + \vartheta \mathbf{G}_{ij}) - \mathbf{k}(\mathbf{T}, \mathbf{D}, \mathbf{Q}) \big) \tag{3.25}$$

for $1 \leq i \leq j \leq 3$. Here, $\mathbf{V}$ denotes the standard basis vector

$$\mathbf{V}_{ij} = (\delta_{ik} \delta_{jl})_{k,l=1}^{3} \tag{3.26}$$

with δ_{ik} being the Kronecker symbol.

Time integration can be accomplished by any integration method that gives sufficiently accurate results, while it does not cause too extensive computational costs. Hügel [41] compared the Euler forward method, Euler backward method, Midpoint method and Runge-Kutta method regarding efficiency and accuracy. As a result, only the Runge-Kutta method fulfilled both requirements for rate independent material models: high accuracy at low computational costs. Fellin and Ostermann [28] proposed a similar second-order method which is derived from the Euler scheme. Additionally, this method provides an error and step size control based on Richardson extrapolation. This method is adopted in this work and will be shortly explained.

We collect all the variables of our problem in one vector

$$\begin{aligned} \mathbf{y} = {} & [T_{11}, T_{22}, T_{33}, T_{12}, T_{13}, T_{23}, B_{1111}, B_{1122}, B_{1133}, B_{1112}, B_{1113}, B_{1123}, \\ & B_{2211} \ldots B_{2323}, Q_1, \ldots, Q_m, (G_{11})_1, \ldots, (G_{23})_m]^T \quad . \end{aligned} \tag{3.27}$$

Denoting the right-hand sides of eq. (3.20) and eq. (3.27) by $\mathbf{F}$, the initial value problem

$$\mathbf{y}'(t) = \mathbf{F}(\mathbf{y}(t)), \quad \mathbf{y}(0) = \mathbf{y}_0 \tag{3.28}$$

has to be solved. Therefore we choose an initial step size τ_0 satisfying $0 \leq \tau_0 \leq \Delta t$ and calculate the two approximations

$$\mathbf{v} = \mathbf{y}_0 + \tau \mathbf{F}(\mathbf{y}_0) \quad , \tag{3.29}$$

$$\mathbf{w} = \mathbf{y}_0 + \frac{\tau}{2}\mathbf{F}(\mathbf{y}_0) + \frac{\tau}{2}\mathbf{F}(\mathbf{y}_0 + \frac{\tau}{2}\mathbf{F}(\mathbf{y}_0)) \quad . \tag{3.30}$$

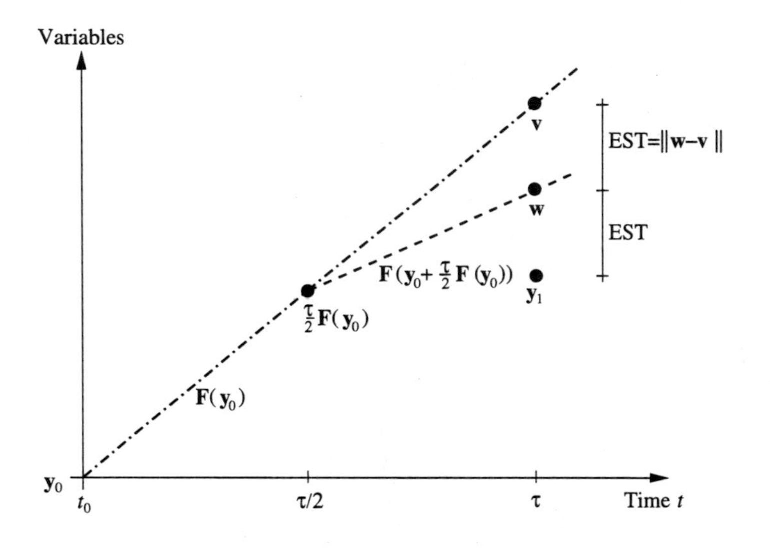

Fig. 3.2: Second-order method for time integration

An estimate for the error – derived via a Taylor series – reads

$$\text{EST} = ||\mathbf{w} - \mathbf{v}|| = \sqrt{\sum_{i=1}^{6} \left(\frac{w_i - v_i}{s_i}\right)^2} \quad . \tag{3.31}$$

with $s_i = \max(y_i, w_i) + \text{ATOL}$.

Using this error estimate, the following step size strategy can be used [35] :

We accept the step, if the error estimate EST is below the user-supplied tolerance TOL. Extrapolating the approximation

$$\mathbf{y}_1 = 2\mathbf{w} - \mathbf{v} \tag{3.32}$$

we obtain a second-order approximation to the solution. Furthermore we can enlarge the step size according to

$$\tau_{new} = \tau \cdot \min\left(5, 0.9\sqrt{\frac{\mathrm{TOL}}{\mathrm{EST}}}\right) \quad . \tag{3.33}$$

If the estimated error EST is larger than TOL, the last step is rejected and redone with a smaller step size given by

$$\tau_{new} = \tau \cdot \max\left(0.2, 0.9\sqrt{\frac{\mathrm{TOL}}{\mathrm{EST}}}\right) \quad . \tag{3.34}$$

Further details of the implementation of the hypoplastic material model can be found in section 3.2.3.

3.2 Constitutive models

3.2.1 Hypoplasticity

A suitable constitutive model is crucial for numerical modelling of material behaviour. As granular materials show a variety of behaviours that are in many ways different from those of other materials, advanced constitutive relations are necessary.

Hypoplasticity is a nonlinear, tensorial equation of the rate type. It can be represented in the following form:

$$\overset{\circ}{\mathbf{T}} = \mathcal{L} : \mathbf{D} + \mathbf{N}\|\mathbf{D}\| \tag{3.35}$$

$\mathcal{L}$ and $\mathbf{N}$ are constitutive tensors. $\mathbf{D}$ is the symmetric part of the velocity gradient. In contrast to Hypoelasticity [98], eq. (3.35) is incrementally nonlinear in stretching $\mathbf{D}$ [49]. $\overset{\circ}{\mathbf{T}}$ is the objective ZAREMBA-JAUMANN stress rate of the Cauchy stress tensor

$$\dot{\mathbf{T}} = \overset{\circ}{\mathbf{T}} + \mathbf{W} \cdot \mathbf{T} - \mathbf{T} \cdot \mathbf{W} \tag{3.36}$$

where $\mathbf{W}$ denotes the spin tensor [99]. In the here used version of von Wolffersdorff [102] $\mathcal{L}$ and $\mathbf{N}$ are functions of the Cauchy stress and the void ratio and read:

$$\mathcal{L} := f_b f_e \frac{1}{\hat{\mathbf{T}} : \hat{\mathbf{T}}} (F^2 \mathcal{I} + a^2 \hat{\mathbf{T}}\hat{\mathbf{T}}), \tag{3.37}$$

$$\mathbf{N} := f_d f_b f_e \frac{Fa}{\hat{\mathbf{T}} : \hat{\mathbf{T}}} (\hat{\mathbf{T}} + \hat{\mathbf{T}}^*), \tag{3.38}$$

$$\text{with} \quad \hat{\mathbf{T}} := \mathbf{T}/\text{tr}\mathbf{T} \quad , \quad \hat{\mathbf{T}}^* := \hat{\mathbf{T}} - \tfrac{1}{3}\mathbf{1}, \tag{3.39}$$

$$a = \frac{\sqrt{3}(3 - \sin\varphi_c)}{2\sqrt{2}\sin\varphi_c} \tag{3.40}$$

$$F := \sqrt{\frac{1}{8}\tan^2\psi + \frac{2 - \tan^2\psi}{2 + \sqrt{2}\tan\psi\cos 3\vartheta}} - \frac{1}{2\sqrt{2}\tan\psi}, \tag{3.41}$$

$$\tan\psi := \sqrt{3}\|\hat{\mathbf{T}}^*\| \quad , \quad \cos 3\theta := -\sqrt{6}\frac{\text{tr}\,(\hat{\mathbf{T}}^* \cdot \hat{\mathbf{T}}^* \cdot \hat{\mathbf{T}}^*)}{(\hat{\mathbf{T}}^* : \hat{\mathbf{T}}^*)^{\frac{3}{2}}} \quad . \tag{3.42}$$

The scalar factors f_d, f_e and f_b are functions of the void ratio e and the mean pressure $p_s := -\text{tr}\,\mathbf{T}/3$ and comprise the dependence on the mean pressure (barotropy) and the density (pyknotropy):

$$f_d := \left(\frac{e - e_d}{e_c - e_d}\right)^{\alpha} \quad ,$$

$$f_e := \left(\frac{e_c}{e}\right)^{\beta} \quad ,$$

$$f_b := \left(\frac{e_{i0}}{e_{c0}}\right)^{\beta} \frac{h_s}{n} \frac{1 + e_i}{e_i} \left(\frac{-\text{tr}\,\hat{\mathbf{T}}}{h_s}\right)^{1-n} \left[3 + a^2 - a\sqrt{3}\left(\frac{e_{i0} - e_{d0}}{e_{c0} - e_{d0}}\right)^{\alpha}\right]^{-1} \quad .$$

e_i, e_c and e_d are the maximum, the critical and minimum void ratios (characteristic void ratios) and depend on the mean pressure. Bauer [14] proposed the following heuristic relation:

$$\frac{e_i}{e_{i0}} = \frac{e_c}{e_{c0}} = \frac{e_d}{e_{d0}} := \exp - \left(\frac{-\text{tr}\,\mathbf{T}}{h_s}\right)^n \quad . \tag{3.43}$$

e_{i0}, e_{c0} and e_{d0} are the void ratios for a stress-free state. This particular formulation enables the definition of a pressure dependent density index:

$$r_e := \frac{e_c - e}{e_c - e_d} \quad . \tag{3.44}$$

$r_e = 1$ represents the densest and $r_e = 0$ the critical state.

The introduced hypoplastic model uses the following eight parameters:

h_s : granular hardness,
n : exponent,
e_{d0}: minimum void ratio for a stress-free state,
e_{c0}: critical void ratio for a stress-free state,
e_{i0}: maximum void ratio for a stress-free state,
φ_c: critical friction angle,
α : exponent,
β : exponent.

All parameters can be determined from simple index and/or element tests (see Tab. 3.1) by the procedure proposed by Herle [38].

material constants	standard test	common range
φ_c	sandpile, direct shear	25°-40°
h_s	oedometric compression	10 - 50000 [MPa]
n	oedometric compression	0.1-0.4
e_{d0}	density index test	0.2-1.0
e_{c0}	density index test	0.4-1.5
e_{i0}	related to e_{c0}	0.5-1.6
α	CID-triaxial test,	
	or direct shear	1.0-4.0
β	oedometric compression	1.0-4.0

Tab. 3.1: Standard tests to obtain hypoplastic parameter and their common range according to Nübel [70].

3.2.2 Intergranular strain

As e.g. Jardine et al. [46] showed, the behaviour of granular soils is incrementally non-linear except for a small elastic range not exceeding approx. 10^{-5}. In this range the stiffness is higher and approximately independent of strain concept. In order to improve the small-strain relationship in hypoplasticity, Niemunis and Herle [73] proposed the intergranular strain.

According to the idea of intergranular strain, the total strain is the sum of deformation of the intergranular layer and of a rearrangement of the skeleton.

The interface deformation is called intergranular strain $\mathbf{S}$ and is considered as a new state variable. When the intergranular strain reaches the limit R, the interface remains deformed while the grains are sliding (Fig. 3.3).

The evolution equation of $\mathbf{S}$ is postulated as:

$$\mathring{\mathbf{S}} = \begin{cases} (\mathcal{I} - \hat{\mathbf{S}}\hat{\mathbf{S}})\rho^{\beta_r} & \text{for } \hat{\mathbf{S}} : \mathbf{D} > 0 \\ \mathbf{D} & \text{for } \hat{\mathbf{S}} : \mathbf{D} \leq 0 \end{cases} , \tag{3.45}$$

where $\mathring{\mathbf{S}}$ is the objective rate of intergranular strain, ρ is defined as the normalised magnitude of $\mathbf{S}$ by $\rho = (\|\mathbf{S}\|)/R)$, R is the magnitude of the intergranular strain and β_r is a material constant.

The stress-strain relationship of Hypoplasticity with intergranular strain reads:

$$\mathring{\mathbf{T}} = \mathcal{M} : \mathbf{D} \quad . \tag{3.46}$$

wherein the fourth order tensor $\mathcal{M}$ represents the incremental stiffness and is calculated from the hypoplastic tensors $\mathcal{L}(\mathbf{T}, e)$ and $\mathbf{N}(\mathbf{T}, e)$. Contrary to standard hypoplasticity, $\mathcal{L}(\mathbf{T}, e)$ and $\mathbf{N}(\mathbf{T}, e)$ are modified (increased) by scalar multipliers including the material constants m_R and m_T. The incremental stiffness is now calculated from

$$\mathcal{M} = [\rho^{\chi} m_T + (1 - \rho^{\chi}) m_R]\mathcal{L} + \begin{cases} \rho^{\chi}(1 - m_t)\mathcal{L} : \hat{\mathbf{S}}\hat{\mathbf{S}} + \rho^{\chi}\mathbf{N}\hat{\mathbf{S}} & \text{for } \hat{\mathbf{S}} : \mathbf{D} > 0 \\ \rho^{\chi}(m_R - m_t)\mathcal{L} : \hat{\mathbf{S}}\hat{\mathbf{S}} & \text{for } \hat{\mathbf{S}} : \mathbf{D} \leq 0 \end{cases} \tag{3.47}$$

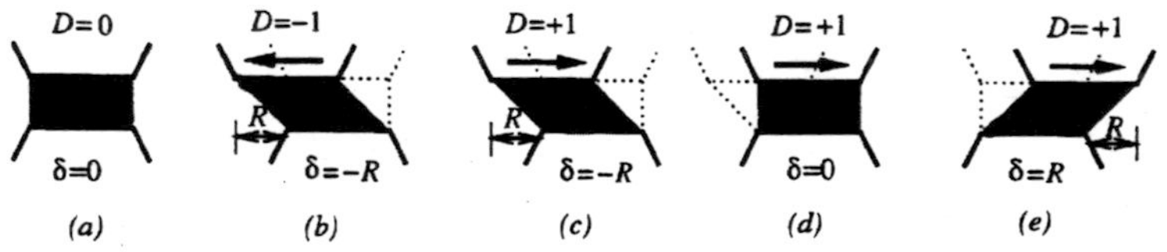

Fig. 3.3: 1D interpretation of the intergranular strain according to Niemunis and Herle [73]: The interface (dark area) is deformed $D = -1$, until the intergranular strain reaches its extremum $\delta = -R$, which cannot be surpassed (b). The interface remains deformed, while the grains are sliding, which corresponds to a rearrangement of the grains. After the reversal of D (c), the deformation concentrates in the interface, until δ vanishes zero (d). Finally δ reaches the limit $\delta = R$ on the opposite side (e).

m_R, m_T, R and β_r and χ are additional material parameters.

The calibration of the extended hypoplastic model with intergranular strains requires biaxial tests or undrained triaxial tests with preceding isotropic compression. Furthermore, small strains have to be measured. As experience shows[2], the typical values (Tab. 3.2) deliver satisfactory results and a more detailed parameter determination usually does not influence the overall results considerably.

R	m_R	m_T	β_r	χ
$1 \cdot 10^{-4}$	5.0	2.0	0.5	6.0

Tab. 3.2: Intergranular strain parameter

3.2.3 How to improve numerical stability

The highly nonlinear character of hypoplasticity leads to severe problems regarding convergence. This is even worse for hypoplasticity with intergranular strain. Besides, inadmissible states exist which should not be reached during calculation.

Several authors of UMATs developed various techniques to overcome these problems. In the following section a brief overview of these techniques will be given and the ones used in this work will be discussed in detail.

3.2.3.1 Stiffness Matrix

Due to the non-linearity, the stiffness matrix can vary dramatically from loading to unloading. This fact may cause slow convergence of the Newton-Raphson iteration and may furthermore result in an abort of the calculation by ABAQUS due to the convergence limits set. To cope with this problem, Tejchman [96] proposes to use the linear terms in $\mathbf{D}$ only to calculate the stiffness matrix. Roddeman [85] uses a central difference scheme to obtain the material stiffness, that is the stiffnesses for loading and unloading are averaged. Both methods ensure an increase of robustness, but one has to be aware, that the loss of quadratic convergence (and therefore increased computational costs) is coming along. In the UMAT used in this work, none of the above techniques has been used.

[2]Personal information of Ivo Herle.

3.2.3.2 Density limit

Extensions of the original hypoplastic model including void ratios or the intergranular strains enabled to solve challenging problems. However, using these extensions brings along a lack of robustness of the numerical calculations [74]. The range of admissible void ratios is limited by e_i and e_d, which both decrease with mean pressure (eq. 3.43), although for some paths a surpassing of the bounds is possible [33].

Niemunis et al. [74] showed, that for standard hypoplasticity the upper bound (e_i) is consistent with the material model, i.e. it will not be violated by the material model. This fact does not hold for intergranular strains, where special zig-zag strain paths can arrive beyond the e_i-limit. Up to now this is an unsolved problem, but as it only appears in very rare cases, it is neglected in this work.

The lower bound e_d is not consistent for both, standard hypoplasticity and hypoplasticity with intergranular strain. Strictly speaking, the calculation must be stopped, whenever the bounds are surpassed. In fact, this is what the implementation programmed by Fellin and Ostermann [29] does. It forces ABAQUS to minimise the increment size, whenever an inadmissible state, e.g. $e < e_d$, is reached. Unfortunately this technique leads to an unstable algorithm, as for certain stress paths a violation of the bounds seems inevitable (see Fig.3.4a).

Niemunis et al. [74] proposed the following modification of the density factor f_d for $e \leq e_d$:

$$\overline{f}_d = -\frac{\sqrt{3}(1+e) + M_T^{(d)} f_b f_e \sqrt{3}(3+a^2)}{M_T^{(d)} f_b f_e 3a} \quad . \tag{3.48}$$

This expression produces results for element tests according to Fig.3.4b but the drastic change of f_d causes convergence problems. The following interpolation rule aims to overcome these problems

$$f_d = \mathrm{sign}(e - e_d)\left(\frac{|e-e_d|}{e_c - e_d}\right)^{\alpha} + \left[1 - \mathrm{sign}(e-e_d)\left(\frac{|e-e_d|}{e_c-e_d}\right)^{\alpha}\right]\overline{f}_d \quad , \tag{3.49}$$

but as element tests show, the interference to the material model is too strong.

Finally, in the UMAT used in this work the limits given by e_d and e_i are ignored in a way, that the density factor f_d is set to zero if $e < e_d$ and it is set to one if $e > e_i$. This means that the void ratio is allowed to surpass the bounds (Fig.3.4c) and "self-healing" properties of hypoplasticity are trusted to solve the problem during the ongoing calculation. The surpassing is reported in a state variable.

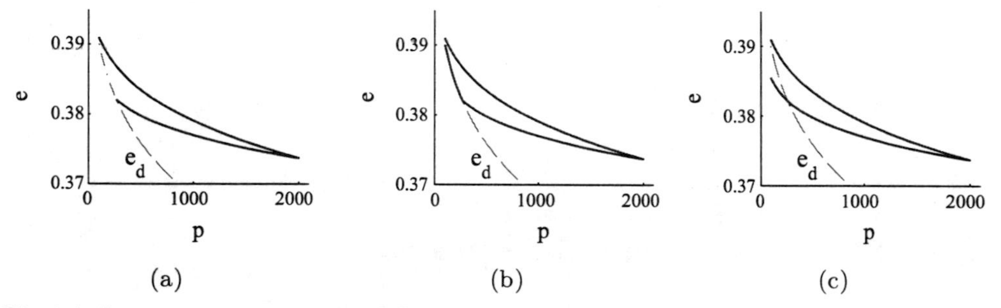

Fig. 3.4: Isotropic compression with Hypoplastic material model, pressure vs. void ratio. (a) Calculation stops, when lower bound e_d is reached; (b) Modification proposed by Niemunis et al. [74]; (c) Modification used in this thesis: lower bound e_d is ignored, violation is reported.

3.2.3.3 Tensile stresses

If we look at the basic hypoplastic constitutive model (eq. 3.35), it is evident, that the stiffness vanishes for $\mathrm{tr}\,\mathbf{T} \to 0$ and one obtains negative stiffness for $\mathrm{tr}\,\mathbf{T} > 0$. This is another cause of numerical instabilities.

Nübel and Niemunis [71] used an elastic nucleus in the stress space around $\mathrm{tr}\,\mathbf{T} = 0$ to overcome this problem. Whenever $\mathrm{tr}\,\mathbf{T} > \mathrm{ELA}$, with ELA being a user-defined threshold, the material response is elastic using the last mean stiffness as Young's modulus and the void ratio is kept constant (Fig.3.5a).

Roddeman [85] just halves the sub step size in his implementation whenever $\mathrm{tr}\,\mathbf{T} \geq 0$. If the sub step size reaches a user defined lower limit, the time integration is stopped. This technique produces a very stable implementation and was slightly improved in the UMAT of Fellin and Ostermann [29], who set the stress rate $\mathring{\mathbf{T}} = 0$ whenever $\mathrm{tr}\,\mathbf{T} > 0$ but continued time integration and guaranteed therefore proper time integration of the remaining state variables (Fig.3.5b), which is only relevant in case of hypoplasticity with intergranular strains. The latter technique was used in this work.

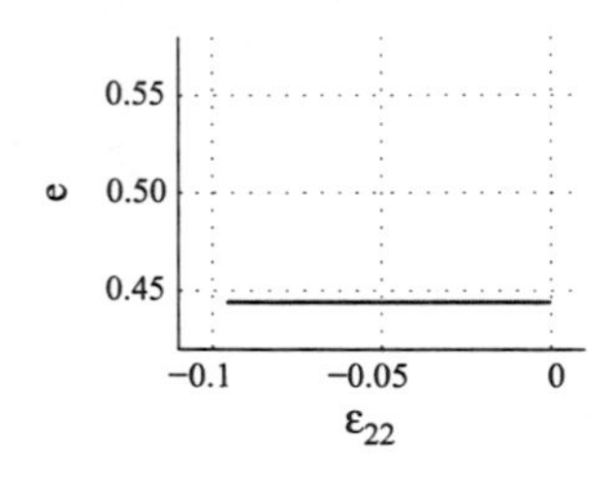

(a) Using elastic nucleus

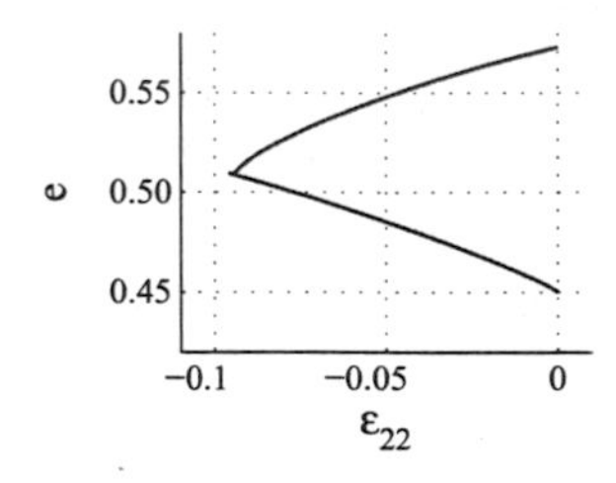

(b) Updated void ratio

Fig. 3.5: Triaxial element test, extension and following compression for low stress level ($p_0 = 0.1$ kPa), strain ε_{22} vs. void ratio e. Note that due to the low stress level dilation arises immediately in (b).

3.2.4 Size of increments and sub steps

The choice of adequate increment and sub step sizes is of paramount importance regarding accuracy and speed of the calculation. For increments used to solve the global equilibrium equation system Kolymbas [50] recommended that every increment should produce a step of equal length in the stress-strain space. For this purpose the following empirical equation can be used:

$$\Delta t = \frac{b|\mathrm{tr}\,\mathbf{T}|}{||\mathring{\mathbf{T}}|| + a||\mathbf{D}||\,|\mathrm{tr}\,\mathbf{T}|} \quad , \tag{3.50}$$

with a and b being empirical constants. Numerical simulations of element tests showed, that $a = 10$ and $b = 0.01$ give sufficiently accurate results. In this work the automatic increment size control of ABAQUS was used.

The second order Euler method used in this thesis for time integration allows an estimation of the error, which can be used to control the sub step size (see section 3.1.4 and appendix B). This is sufficient in case of standard hypoplasticity, but when intergranular strains are used, additional conditions are required to assure numerical stability.

The following additional limits provide numerical stability in case of intergranular strains: The strain increments $\Delta\varepsilon = ||\mathbf{D}||\Delta t$ are limited by

$$\Delta\varepsilon \leq \Delta\varepsilon_{\mathrm{max}} \quad , \tag{3.51}$$

for $\hat{\mathbf{S}} : \mathbf{D} > 0$, where a value of $\varepsilon_{max} \approx 0.0001 \approx R$ is chosen for the computations in this work. For $\hat{\mathbf{S}} : \mathbf{D} \leq 0$ (i.e. change of loading path) a maximum step size of

$$\Delta\varepsilon = \frac{0.1R}{1 - \rho_r^{\beta}} \tag{3.52}$$

is chosen according to Niemunis et al. [74].

3.3 Requirements on constitutive law

Case histories (section 2.2.1) and sophisticated FE calculations (section 2.5.5) indicate that the ground in the vicinity of a tunnel driven by a shield undergoes loading-unloading cycle. Thus, if FE analyses are used to estimate soil deformations due to shield tunnelling, the constitutive model must be able to reproduce the volumetric behaviour of soil under a loading-unloading cycle thoroughly.

3.3.1 Behaviour of soil at cyclic loading

We first consider the stress path of a soil element above the crown (Fig. 3.6), since it is assumed, that the stress path in this area has the biggest impact on settlements. Lee and Rowe [60] investigated this stress path by means of a 3D FE-analysis. They concluded, that the trend of the stress path is similar to that of triaxial compression for a soil element ahead of the tunnel face. Once the face advances beyond the point of interest, the stress path quickly changes to one closer to triaxial extension. This stress path can be simulated by means of laboratory tests, i.e. triaxial or biaxial (= plane deformation). Fig. (3.7) shows some typical results.

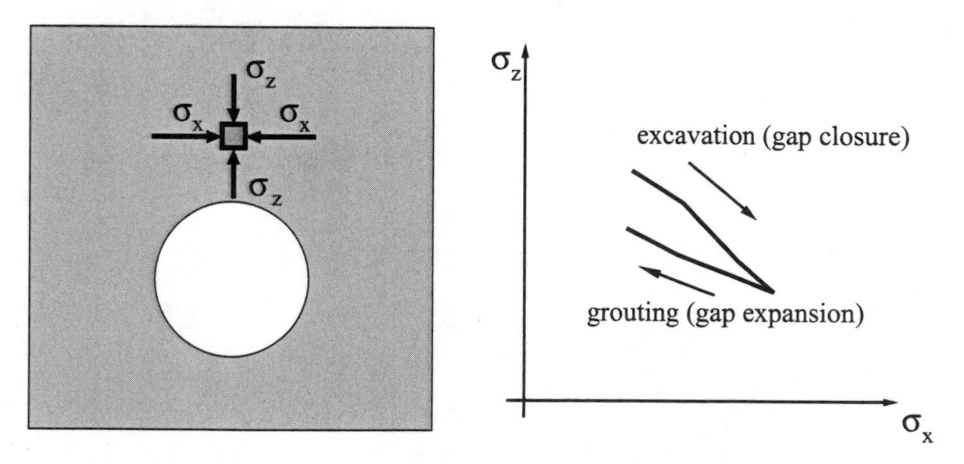

Fig. 3.6: Stress path above the crown due to tunnel excavation and subsequent tail gap grouting [52]

Kolymbas et al. [52] pointed out, that one should pay attention to the dilatancy behaviour of sand (and soil in general). During (triaxial) loading dense sand exhibits first contractancy and later dilatancy. As known, dilatancy is the increase of volume (and, consequently, also of void ratio) that occurs during shearing. Of paramount importance is what happens at reversals of the deformation, i.e. at loading-unloading transitions. Unloading is characterised by a much higher stiffness. Considering the evolution of volumetric strain, it is observed that the previous dilatancy is reversed to contractancy, the rate of which is much higher than the one of dilatancy (Fig. 3.7b). This fact implies the following result: A loading-unloading cycle leaves behind a compaction of the soil [52].

Interestingly, the previously described fundamental behaviour of soil at cyclic loading is not modelled by most of the constitutive equations implemented in standard FE codes. E.g., the so-called Mohr-Coulomb elasto-plastic constitutive equation predicts dilatancy at loading and also immediately after the reversal from loading to unloading (see Fig. 3.8), which is fundamentally wrong. As a result, the Mohr-Coulomb constitutive law is unable to simulate the behaviour of soil in a shield tunnelling problem [68]. In contrast, the hypoplastic constitutive equation reproduces this effect realistically (Fig. 3.9).

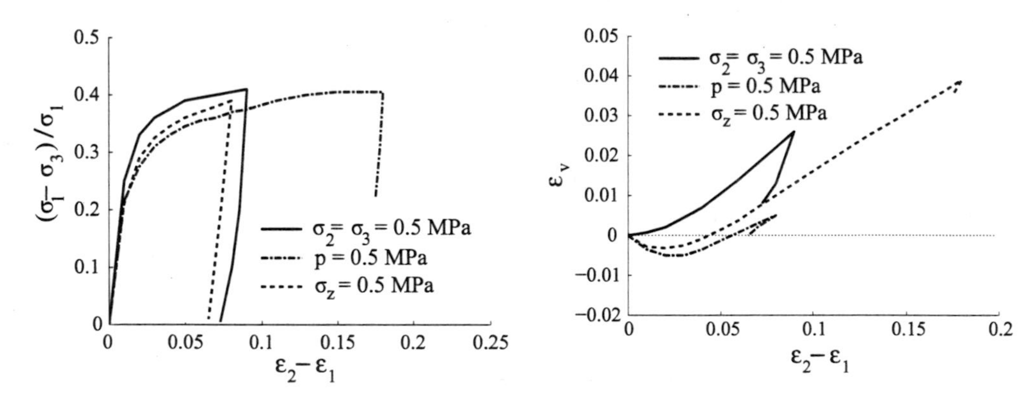

a Deviatoric stress vs. deviatoric strain b Volumetric strain vs. deviatoric strain

Fig. 3.7: Triaxial experiments under different stress paths with Hostun Sand (Lanier et al. [57]). Note the pronounced contractancy upon unloading

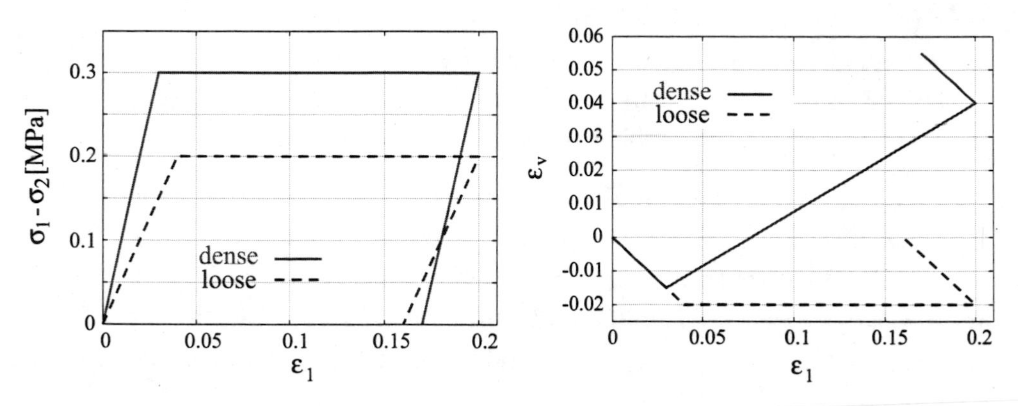

(a) Deviatoric stress vs. axial strain (b) Volumetric strain vs. axial strain

Fig. 3.8: Simulation of triaxial test results with the MOHR-COULOMB constitutive model

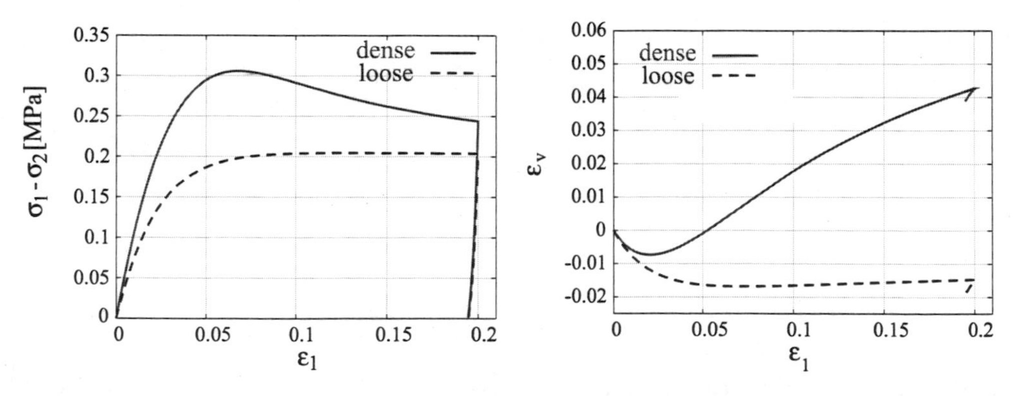

(a) Deviatoric stress vs. axial strain (b) Volumetric strain vs. axial strain

Fig. 3.9: Simulation of triaxial test results with HYPOPLASTICITY.

3.3.2 Boundary conditions

To test the appropriateness of several constitutive models for shield tunnelling analyses, a simple FE-model has been investigated. This model is not intended to simulate the entire soil-shield interaction nor to predict the behaviour of the lining, but merely to show the importance of a realistic constitutive law for the considered problem of gap grouting.

The boundary conditions and the mesh have been chosen as illustrated in Fig. (3.10). The sequence of the applied deformation is shown in Figs. (3.11a) to (3.11c). We start from a circular tunnel with 5 m radius (Fig. 3.11a) and apply a convergence of 7 cm (Fig. 3.11b) to simulate the excavation. In the second step the tunnel cavity is expanded to its original size as displayed in Fig. (3.11c).

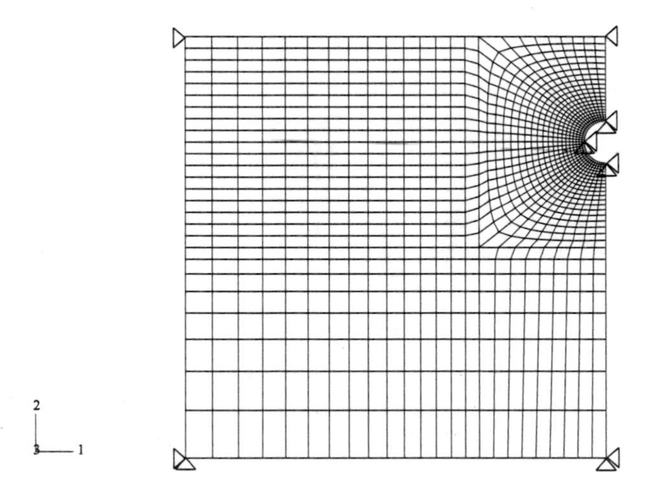

Fig. 3.10: Boundary conditions

3.3.3 Parameters of constitutive models

Triaxial and oedometer tests with Hochstetten sand from the vicinity of Karlsruhe [102] were used for the determination of the material parameters of the hypoplastic model [40]. The initial relative density based on the void ratio has been chosen as $I_e = \frac{e - e_{\min}}{e_{\max} - e_{\min}} = 0.5$.

The parameters of the Mohr-Coulomb model were adjusted as a best fit to the results of the triaxial experiments at $\sigma_0 = 0.1$ MPa, obtaining Young's

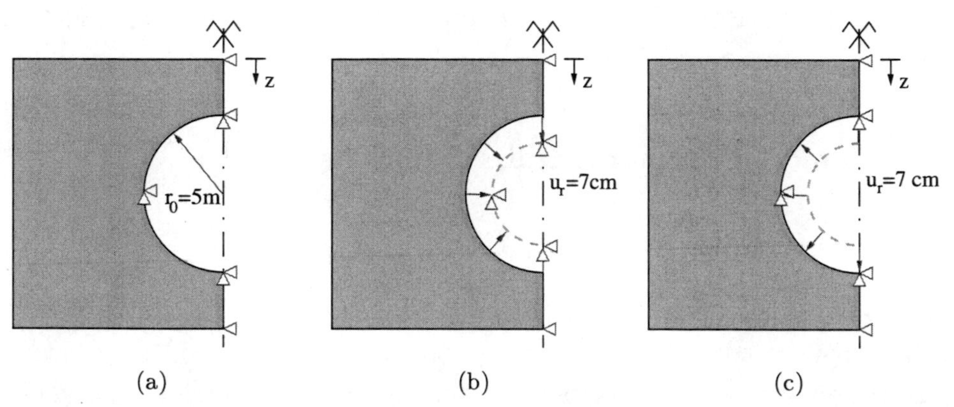

(a) (b) (c)

Fig. 3.11: Displacements applied to the tunnel contour line. (a) Initial configuration; (b) convergence; (c) expansion.

modulus $E_0 = 22$ MPa, friction angle $\varphi = 33\,^\circ$, POISSON's ratio $\nu = 0.3$ and an angle of dilatancy $\psi = 30\,^\circ$.

The Young's modulus E was varied with depth z according to a pressure-dependent law proposed by Ohde [77], with α set to 0.7:

$$E(z) = E_0 \left(\frac{\sigma_z(z)}{\sigma_0} \right)^{\alpha} \tag{3.53}$$

The elasto-plastic and hypoplastic parameters are summarised in Tabs. 3.3 and 3.4.

ϕ [°]	E_0 [MPa]	ν	ψ [°]
33	58	0.3	30

Tab. 3.3: Parameters of the MOHR-COULOMB model for Hochstetten sand.

ϕ_c [°]	h_s [MPa]	n	e_{d0}	e_{c0}	e_{i0}	α	β
33	1000	0.25	0.55	0.95	1.05	0.25	1.50

Tab. 3.4: Parameters of the HYPOPLASTIC model for Hochstetten sand.

The analyses have been carried out with the FE program ABAQUS. Mohr-Coulomb law is a standard part of this FE-package. For the analyses in this section standard Hypoplasticity, version of von Wolffersdorff, was used. For this purpose the UMAT from Fellin and Ostermann [28] has been used, which provides a consistent tangent operator including error control.[3]

[3]The UMAT is available on http://www.geotechnik.uibk.ac.at.

Three calculations with Mohr-Coulomb model with varying dilatancy angle ψ have been performed, to analyse its influence on the results.

3.3.4 Results of the numerical simulations

3.3.4.1 Hypoplasticity

Fig. (3.12) shows the evolution of the settlement trough. One can clearly see that the trough is not annihilated by the cavity expansion — in agreement with field observations [45]. This behaviour is also reflected in the evolution of vertical displacements along the vertical symmetry axis in Fig. (3.13). There, a loosening zone about 2 m above the crown can be identified. It is below the soil arch created during the tunnel convergence. This loosening zone is also shown in a contour plot of void ratios at the end of the convergence-expansion cycle (Fig. 3.14). The volume of settlement trough still increases during gap expansion (Fig. 3.17), due to densification of the soil sideways to the tunnel.

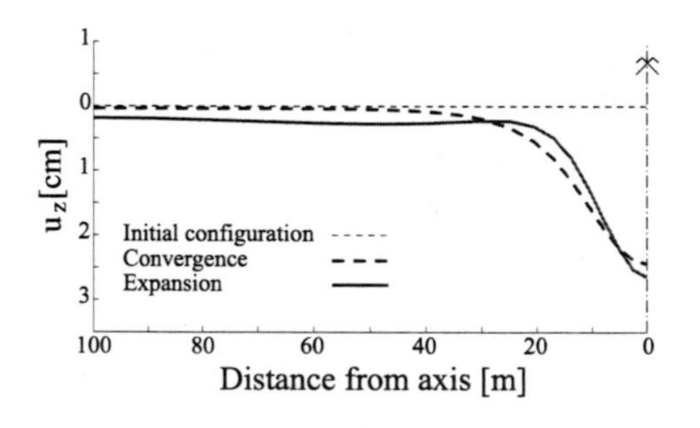

Fig. 3.12: Evolution of the settlement trough, HYPOPLASTICITY.

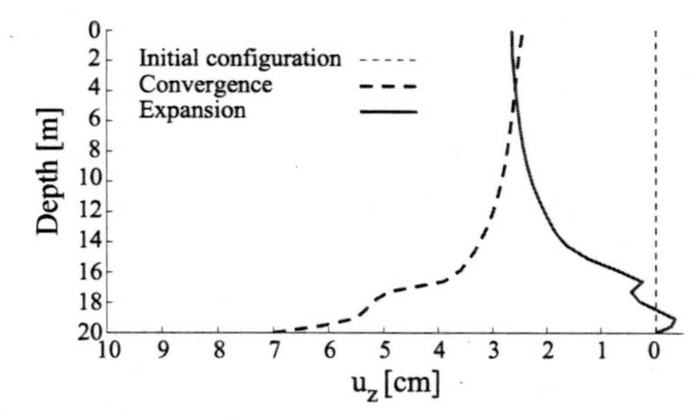

Fig. 3.13: Evolution of vertical displacements u_z along the vertical symmetry axis, HYPOPLASTICITY.

3.3.4.2 Mohr-Coulomb

In contrast to the settlement troughs calculated with hypoplasticity, the volume of settlement trough obtained with the Mohr-Coulomb model (Fig. 3.15) is nearly annihilated by the cavity expansion. Moreover, about 30 m beside the symmetry axis a heave can be observed. This is also depicted in Fig. (3.17), where the volume of settlement trough is plotted against tunnel convergence. While the volume of settlement trough computed with the hypoplastic material law is still increasing during gap expansion, the volume of

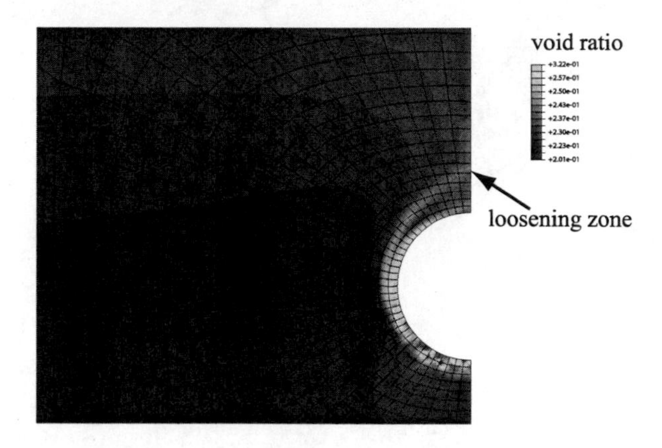

Fig. 3.14: Distribution of void ratio, numerically obtained with HYPOPLASTICITY.

settlement trough computed with Mohr-Coulomb decreases during expansion. In fact — depending on the angle of dilatancy — one even gets a net volume increase.

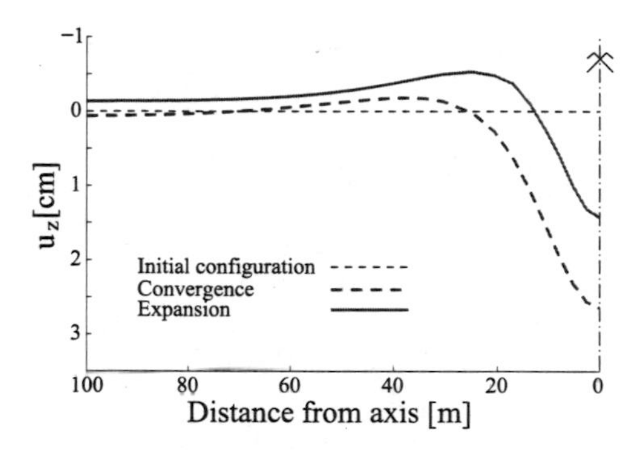

Fig. 3.15: Evolution of the settlement trough, MOHR-COULOMB.

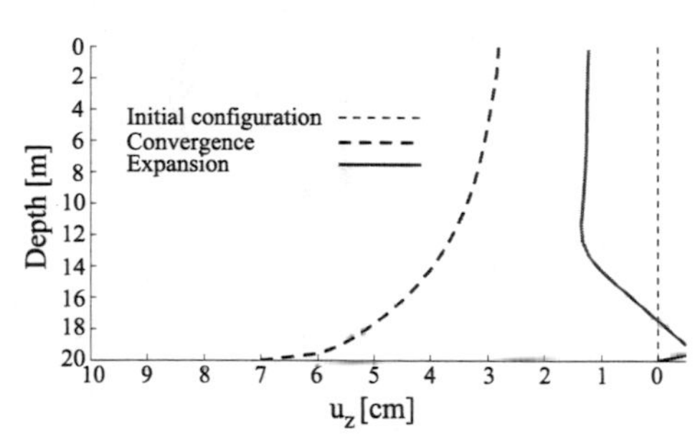

Fig. 3.16: Evolution of vertical displacements u_z along the vertical symmetry axis, MOHR-COULOMB law, $\psi = 30°$.

3.3.5 Conclusion

The settlement due to tail gap closure can not be completely revoked with grouting from the very beginning — apart from excessive grouting. This is due to the contractancy of soil at loading-unloading cycles and has been simulated with finite elements and the hypoplastic constitutive equation. The comparison of the settlement trough obtained using elasto-plasticity and the one obtained with hypoplasticity points out to the advantage of using realistic constitutive laws such as hypoplasticity. In FE models of shield tunnelling

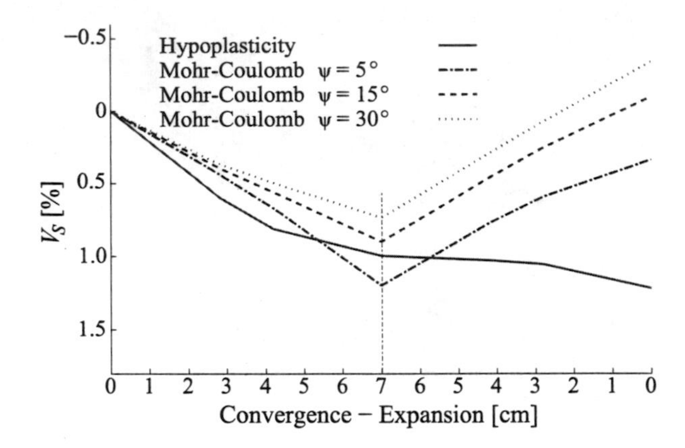

Fig. 3.17: Evolution of volume of settlement trough V_S with the change of the cavity radius.

with advanced boundary conditions (e.g. Kasper and Meschke [47] or Komiya [53]) loading-unloading cycles are produced. Thus, the use of a realistic constitutive model is of major importance.

Chapter 4

Own Tests

While there are a lot of model tests regarding shield tunnelling in clay, only few can be found for sand or gravel, and there is no model known to the author that takes tail gap grouting into account. The model test presented in this thesis has been developed by the author to simulate shield tunnelling in sand. The main emphasis is put on the influence of tail gap grouting on soil displacements.

4.1 Test apparatus

The test apparatus consists of a container of 100 cm width, 40 cm depth and 100 cm height and a cylinder of 40 cm length and a diameter of 17 cm (see Fig. 4.1). The front wall of the container is made of glass (1) to enable taking photographs, the 2 cm thick side walls and the 3 cm thick back wall are made of chip boards (2). The frame is made of L-sections from aluminum.

The width of the container has been chosen according to the following consideration: to reduce the influence of the side walls, the settlements at the side walls, which are placed 50 cm from the axis, must be negligible. The settlement trough is assessed using eq. (2.8). The standard deviation i of the GAUSS-curve (i.e. settlement trough) then reads:

$$i = \frac{D}{2}\left(\frac{H}{D}\right)^{0.8} = \frac{17.0\ \text{cm}}{2}\left(\frac{42.5\ \text{cm}}{17.0\ \text{cm}}\right)^{0.8} \approx 17.7\ \text{cm} \quad . \tag{4.1}$$

Thus, the settlements in a distance of $3i = 3 \cdot 17.7$ cm $= 53.1$ cm equals 1% of the maximum settlement above the crown and, consequently, the influence of the side walls can be neglected.

A cylinder with variable diameter from 16 cm to 17 cm is placed in the centre line of the container at a height of 58 cm. It is supported by thin

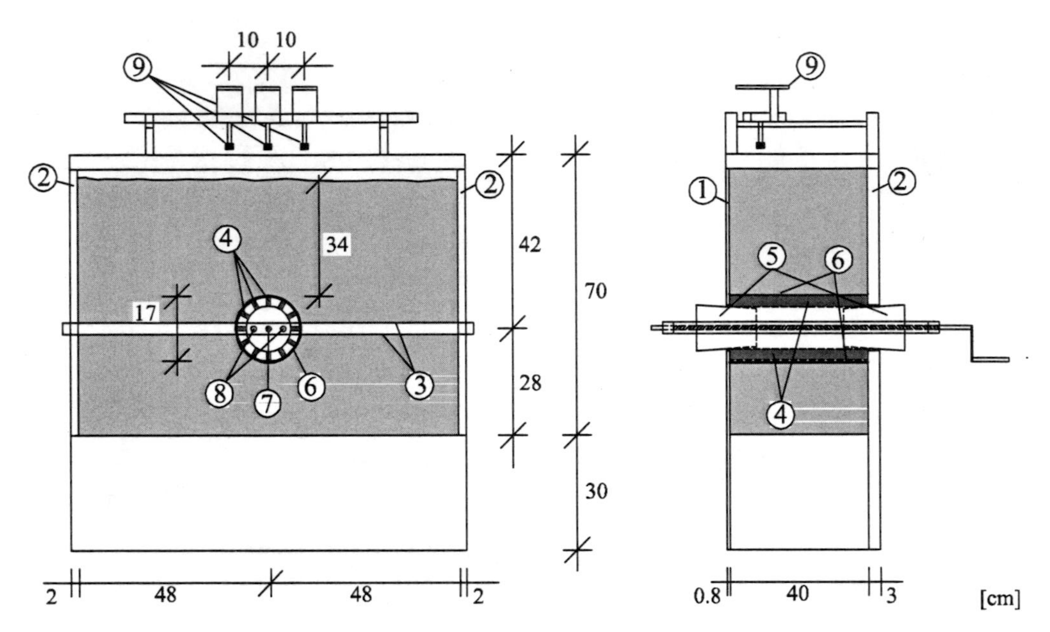

Fig. 4.1: Sketch of the model container: (1) Glass panel; (2) chip boards; (3) aluminium beams; (4) cylinder segments (sheet metal and 3 cm high plastic beams); (5) aluminium frustums; (6) rubber membrane; (7) screw thread; (8) support rods to hinder twisting of the frustum; (9) infrared sensors.

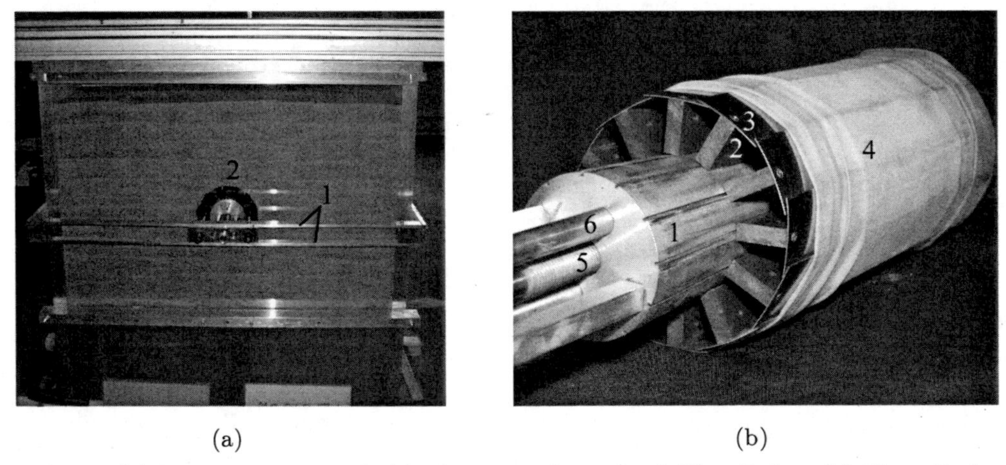

(a) (b)

Fig. 4.2: (a) Test apparatus with (1) aluminium beams and (2) cylinder. (b) The cylinder consists of (1) two frustums, 12 segments composed of (2) plastic beams and (3) metal sheets, (4) rubber membrane, (5) screw thread and (6) rods against twisting.

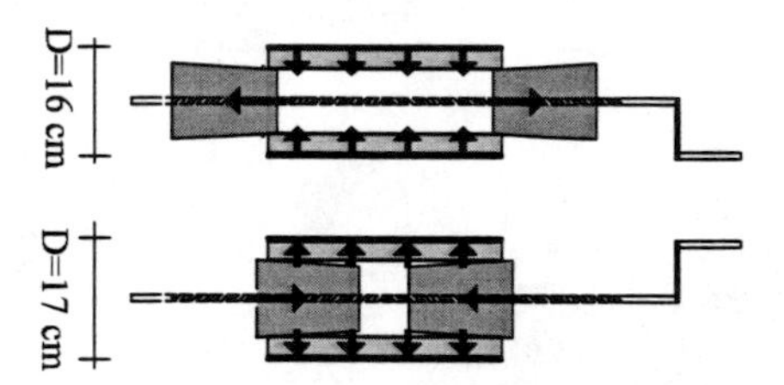

Fig. 4.3: (1) Frustums are moved outward and the segments move towards the axis. The diameter decreases. (2) Frustums are moved inward and the segments outward. The diameter increases.

aluminium beams (see Fig. 4.1,(3)), which allow vertical movements of the cylinder axis [1]. The cylinder consists of 12 segments (4), which are supported by two aluminium frustums (5). A rubber membrane coats the cylinder (6). The segments are made from sheet metal and 3 cm high plastic beams. The frustums are guided by a screw thread (7), whereby the frustums can be moved along the tunnel axis. Two additional rods (8) save the frustums against twisting. The movement of the frustum along the screw threads leads to a gradual decrease or increase of the cylinder diameter (Fig. 4.3).

After the sand is built in, shield tunnelling is modelled in two steps. In a first step the cylinder is contracted by ΔD to simulate face loss and shield loss (Fig. 4.4). In the second step the cylinder is dilated by the same amount to simulate tail gap grouting.

All tests have been carried out under normal gravitational conditions, so called $1g$ tests. Due to the stress dependence (barotropy) of soil behaviour, these tests have limited applicability: The main problem of $1g$ tests is the high dilatancy, high friction angle and the low stiffness of the soil at low pressures. To overcome this shortcoming, test series using Soiltron have been carried out. Soiltron is soil enriched with light and soft additives and behaves in the model test as prototype soil under higher stresses [58]. The results from these tests may be found in appendix C.

Apart from the limitations of barotropy [2] the so-called particle size effect is a further limitation of geometrically scaled tests. As the grain size of the model soil is in the range of natural soils, the shear bands, which generally have an approximate thickness of 10 times the mean particle size, will have the same thickness in the model as in reality. The particle size effect is reported by

[1] As shield and tunnel are only supported by the surrounding soil in reality, a compliant bearing has been chosen.

[2] The stress dependence of soil behaviour is termed barotropy.

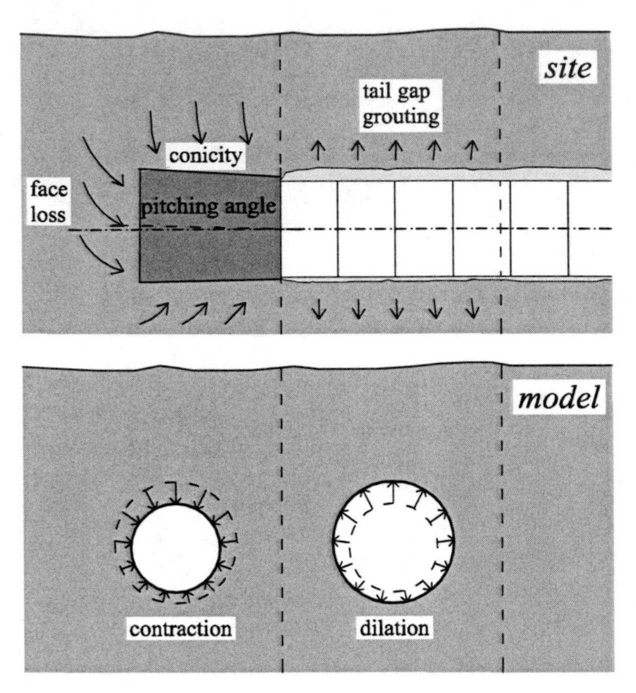

Fig. 4.4: Shield tunnelling modelled in 2 steps: Face loss and shield loss is simulated by contraction of the cylinder, tail gap grouting by dilation of the cylinder.

Cha [21] to play a minor role, if the relevant dimension of the model (e.g. the diameter $D = 170$ mm) is > 30 times of the mean diameter d_{50} of the soil model ($d_{50} = 0.58$ mm), which is the case in this work.

4.1.1 Measurement of soil displacement by Particle Image Velocimetry

Particle Image Velocimetry (PIV) is a velocity (and displacement) measuring technique, that was originally developed in the field of experimental fluid mechanics and is recently increasingly applied in geotechnical modelling [30, 72, 84, 104]. Digital images are taken for the evaluation of a two-dimensional field of a granular body. A 5 Megapixel camera Olympus Dimage 7i has been used placed in a distance of 70 cm to the container with the camera axis perpendicular to the investigated front.

After each test the digital images are transferred to a computer. Each picture is divided in so called interrogation cells (Fig. 4.6). By means of cross correlation a local displacement vector is determined for each interrogation cell between two consecutive digital images:

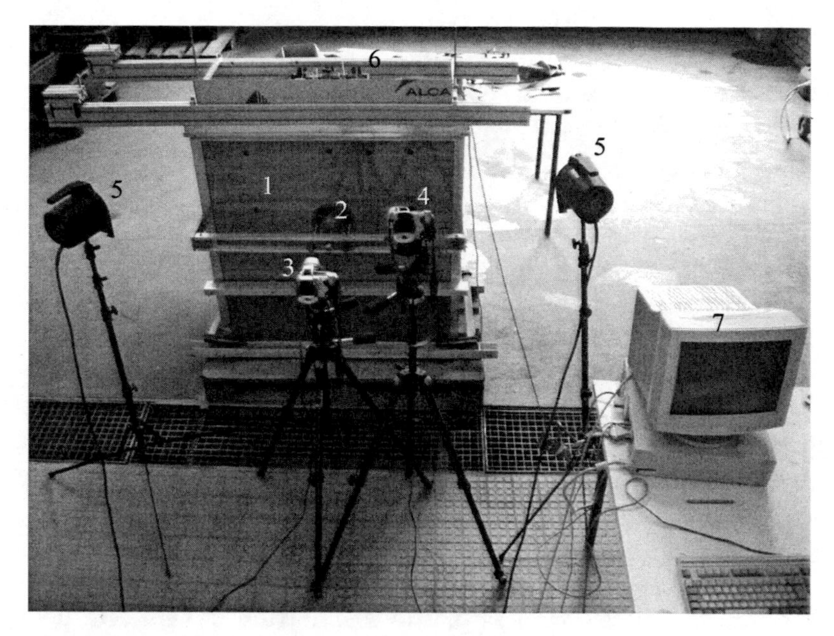

Fig. 4.5: Test setup: (1) Glas panel; (2) cylinder with variable diameter; (3+4) digital cameras; (5) halogen lamp for illumination; (6) infrared measurement device; (7) PC for data aquisition.

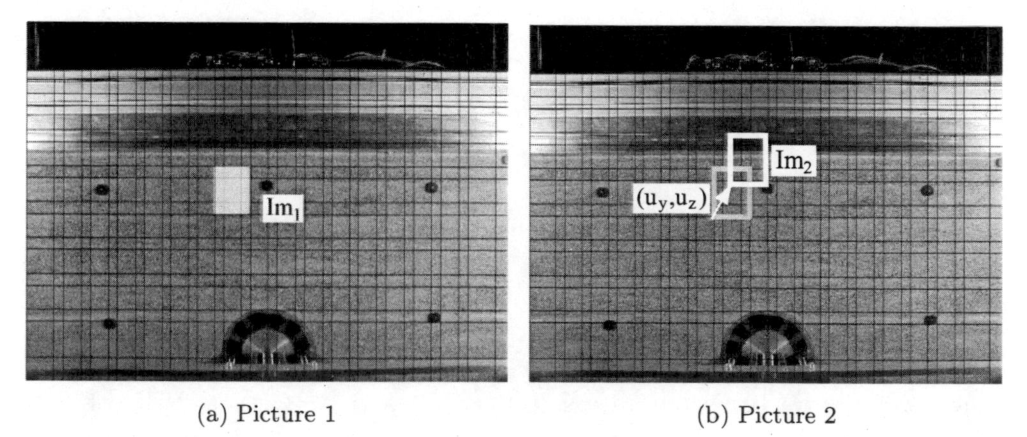

(a) Picture 1 (b) Picture 2

Fig. 4.6: Interrogation cells of consecutive pictures. The lines show the border lines of the overlapped interrogation cells with medium cell size.

$$\mathbf{R}(u,v) = \sum_{i=-M/2}^{M/2} \sum_{j=N/2}^{N/2} \mathrm{Im}_1(i,j)\mathrm{Im}_2(i+u_y, j+u_z) \tag{4.2}$$

$\mathrm{Im}_1(i,j)$ ... image intensity of pixel (i,j) of interrogation cell Im_1 in picture 1.
$\mathrm{Im}_2(i,j)$ ... image intensity of pixel $(i+u, j+v)$ of interrogation cell Im_2 in picture
M ... width of interrogation cell (in pixel),
N ... length of interrogation cell (in pixel),
R ... cross correlation coefficient.

The array $\mathbf{R}$ gives the correlation value for all displacement vectors (u_y, u_z) between the two interrogation cells [82]. The correlation value measures the degree of match between two interrogation cells. Hence, the coordinates of the peak correlation in the correlation array (see Fig. 4.7) gives the displacement vector of the interrogation cell.

One of the most crucial features of PIV evaluation is that the peak of the correlation can be measured to sub pixel accuracy. Since the input data itself is discretised, the correlation values exist only for integral shifts. Interpolation of the data yields accuracies on the order of 1/10th [82] to 1/40th [103] of a pixel.

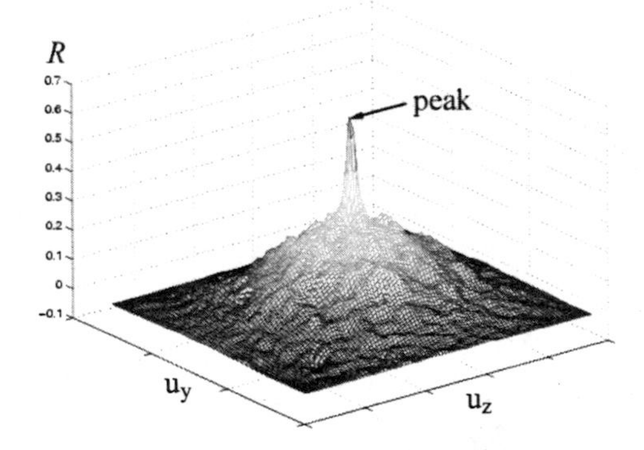

Fig. 4.7: Correlation plane with correlation peak.

The analyses are performed using MatPIV[3], a free available PIV-software running with Matlab. An adaptive multipass method is applied, where the initial cell size of 98 by 128 pixels is reduced in 4 passes to a cell size of 12 by 16 pixels (see Fig. 4.8). Thereby, the vector field computed with the first cell

[3]MatPIV can be obtained from: http:\\ www.math.uio.no\˜jks\matpiv\

size is used as reference vector field for the next pass. In the next evaluation, the cell size is half the previous size and the vector calculated in the first pass is used as best choice cell shift for the second evaluation.

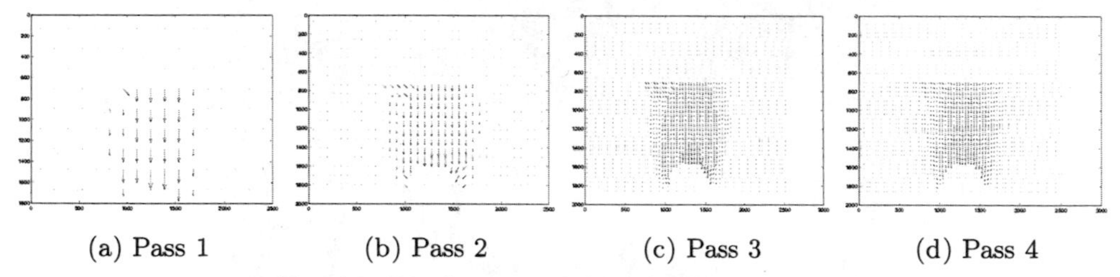

(a) Pass 1 (b) Pass 2 (c) Pass 3 (d) Pass 4

Fig. 4.8: Displacement fields of different passes.

The pictures taken from the model cover an area of 71.9 cm by 52.7 cm. The resolution of the camera is 2560 by 1920 pixels and the medium grain size of Ottendorf-Okrilla sand is $d_{50} = 0.58$ mm. Thus, the picture resolution is 0.29 mm/pixel and 0.5 d_{50}/pixel, respectively. Approximately 48 sand grains (6 by 8) are covered by the smallest interrogation cells, which satisfies the requirement of a minimum of five particles reported by Raffel et al. [82] by far. Moreover, the interrogation cells are smaller then the expected shear bands width, which may be expected in the range from 10 times d_{50} [86] to 16 times d_{50} [100]. Therefore, position and extent of shear bands can by evaluated with sufficient accuracy to the authors opinion.

The strains are calculated from the displacement field by the method of least squares:

$$\begin{aligned} \varepsilon_{yy} &= \left(\frac{\mathrm{d}u_y}{\mathrm{d}y}\right) \approx \frac{2u_{y,i-2} + u_{y,i-1} - u_{y,i+1} - 2u_{y,i+2}}{10\Delta x} \\ \varepsilon_{zz} &= \left(\frac{\mathrm{d}u_z}{\mathrm{d}z}\right) \approx \frac{2u_{z,i-2} + u_{z,i-1} - u_{z,i+1} - 2u_{z,i+2}}{10\Delta y} \quad . \end{aligned}$$

The volumetric strains are obtained from:

$$\varepsilon_v = \varepsilon_{yy} + \varepsilon_{zz} \quad .$$

4.1.1.1 Measurement of settlements with PIV

PIV was also used to measure settlements. For this purpose the displacements from the interrogation cells that cover the area below the surface were assumed to represent the surface settlement (see Fig. 4.9).

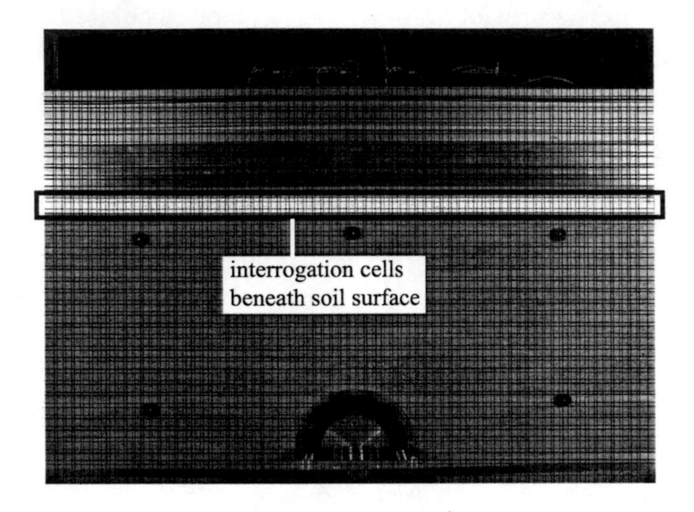

Fig. 4.9: Interrogation cells used to evaluate surface settlements.

4.1.1.2 Accuracy of PIV

The accuracy of the PIV method depends on many factors, such as density of particles, distance of the camera to the specimen, quality of the camera lens, resolution of the camera or quality of the lightning. Fromm [30] scrutinised the accuracy of PIV with the proposed setup by a special test.

For this purpose, sand was fixed on a metal plate (see Fig. 4.10). The plate was rigidly coupled to a wagon with a rod and placed behind the glass panel. The wagon was displaced for 0.5 mm. A dial gauge measured the displacements of the wagon and hence, of the metal plate. Additionally photographs have been taken and analysed with PIV. The comparison of the PIV results with the gauge measurement showed that an accuracy of 5 μm is achieved, which is sufficient for the measurement of soil displacements.

4.2 Test programme

Four test series à three tests have been carried out[4]. Initial density and the change of diameter ΔD have been varied. The overburden has been fixed to

[4] The tests have been carried out by Jana Bochert and Andrew Ross.

Fig. 4.10: Device to control accuracy of test setup. (1) Metal plate with applied sand; (2) wagon; (3) dial gauge.

34 cm $= 2D$.

4.2.1 Diameter change

The tunnel diameter of 17 cm has been decreased by 0.5 cm, 0.66 cm or 1.0 cm. Assuming a geometric scale of 1/50 this corresponds to a tunnel of 8.50 m diameter having a tail gap of 12.5 cm, 16.5 cm or 25 cm, respectively, and 17 m overburden.

The changes have been applied in increments of 0.16 cm to keep the test series comparable. Therefore three, four or six increments were necessary to reach the desired gap.

4.2.2 Material properties

A commercially available quartz sand from a deposit in Ottendorf-Okrilla in Germany[5] has been used. This sand, as well as Soiltron, has been extensively tested by Laudahn [58], most of the soil parameters are taken from this work.

4.2.3 Initial density

A homogeneous initial density is of major importance to guarantee reproducibility of the tests. The methods to build in dry non-cohesive soil are hand tamping, fluidisation and pluviation. Hand tamping means to build in

[5] The sand is purchased from Euroquarz GmbH, D-01458 Ottendorf-Okrilla.

Tab. 4.1: Properties of Ottendorf-Okrilla sand [58]

Mean grain size	d_{50}	0.58 mm
Coefficient of uniformity	$U = d_{60}/d_{10}$	3.6
Particle shape		angular to sub angular
Mineral composition		>97% quartz
Specific gravity	γ_s	2.635 g/cm^3
Max. void ratio	$e_{\max}$	0.75
Min. void ratio	$e_{\min}$	0.42

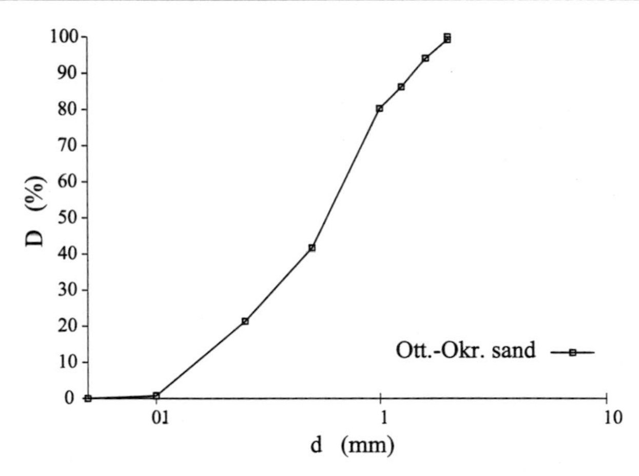

Fig. 4.11: Grain size distribution curve [58]

layers of predefined thickness and after that compacting by tamping. Fluidisation means that a container filled with sand is filled with water by an upward flow, which is stopped when the water level reaches the surface. After the sand has settled down, the specimen is dewatered and dried and the test may start. Pluviation means that sand is rained over the soil model from a specified height of fall. As fluidisation produces very loose samples and pluviation causes segregation of Soiltron, hand tamping was used to prepare the model.

In order to produce a “loose” initial state, the sand was poured in layers of 5 cm with 25 cm height of fall. Then a rake was drawn trough the layer to guarantee an uniform and reproducible state.

In the second method the sand has been poured in the same way as above,

Fig. 4.12: Soil layers of model tests, “dense” initial state.

but every layer has been tamped. This method produced a “dense” initial state. In this method segregation took place, as can be seen in Fig. (4.12).

To assure a constant distribution of the density across the model, the undercompaction method has been used in test series **OT0.66** and **OR0.5** [15]. This method takes into account the densification of the layer due to the additional load of further layers. Using this method, a density distribution with constant relative density $I_e = \frac{e_{\max}-e}{e_{\max}-e_{\min}}$ is produced, while the previous method produces a density distribution with constant relative (pressure dependent) density $r_e = \frac{e-e_d}{e_c-e_d}$ according to the constitutive model of hypoplasticity. In the author's opinion neglecting undercompaction is more realistic. First, in the framework of hypoplasticity a soil with the same r_e at different pressures has similar mechanical behaviour. Second it is closer to natural conditions. In geological sedimentation processes the sand is deposited with constant density. With ongoing sedimentation the lower layers densify due to increasing pressure. The density distribution arising from this process corresponds to a distribution with constant relative density r_e.

4.3 Results

Looking at the qualitative behaviour of soil, i.e. deformation patterns and shear bands, the tests can be divided into two classes: tests with low initial density (“loose” tests) and test with high initial density (“dense” tests). It is important to note, that “loose” and “dense” are not strictly related to the index of relative density (Fig. 4.13). They rather indicate how the soil behaves in the tests.

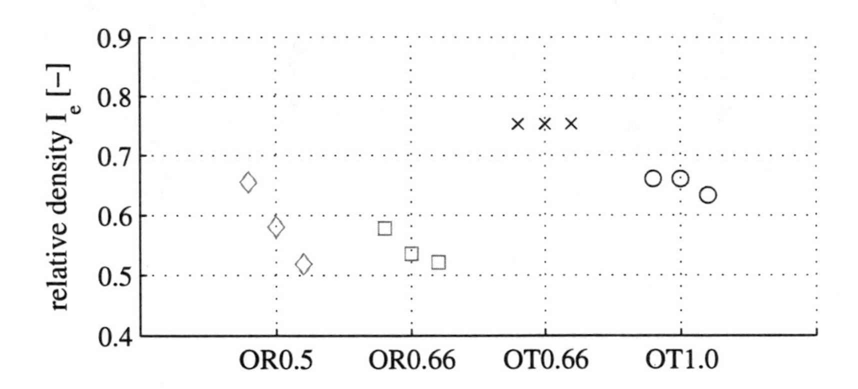

Fig. 4.13: Initial densities

Test No.	void ratio e	rel. density index I_e	pre- paration	material	change of diameter ΔD	in- crements
OR0.50-1	0.53	0.65	raked	O.O.	0.50	3
OR0.50-2	0.56	0.58	raked	O.O.	0.50	3
OR0.50-3	0.58	0.52	raked	O.O.	0.50	3
OR0.66-1	0.56	0.58	raked	O.O.	0.66	4
OR0.66-2	0.57	0.54	raked	O.O.	0.66	4
OR0.66-3	0.58	0.52	raked	O.O.	0.66	4
OT0.66-1	0.50	0.75	tamped	O.O.	0.66	4
OT0.66-2	0.50	0.75	tamped	O.O.	0.66	4
OT0.66-3	0.50	0.75	tamped	O.O.	0.66	4
OT1.0-1	0.53	0.66	tamped	O.O.	1.00	6
OT1.0-2	0.53	0.66	tamped	O.O.	1.00	6
OT1.0-3	0.54	0.63	tamped	O.O.	1.00	6

Tab. 4.2: Initial conditions and soil behaviour of all tests.

The qualitative behaviour of soil is discussed by means of two representative tests, a "loose" and "dense" one. Then a summary of the test results is given and the influence of the diameter change is discussed in detail.

4.3.1 "Loose" initial state

Test **OR066-3** has been chosen to represent the tests with "loose" sand. This test belongs to the test series **OR0.66** with a change of diameter of 0.66 cm and pure sand. Its initial density index was $I_e = 0.52$. In test **OR066-3** the dilation/contraction of the tube was applied in 4 increments.

4.3.1.1 Contraction

At the beginning of the model test the incremental displacements are concentrated in a trapezoidal zone above the crown, which is shown in Fig. (4.16a). The displacement vectors point towards the tunnel axis. During the ongoing contraction, the trapezoidal zone, where incremental displacements occur, becomes narrower.

This narrowing of the area where displacements occur can be seen in Figs. (4.18a) to (4.18d), where the incremental strain in vertical direction ε_{22} is plotted. The shear bands, which enclose the zone of displacement, become steeper with ongoing contraction. Starting from an angle of approx. 45° to the horizontal, the angle increases up to 90° until a shaft is formed. After the formation of the shaft, soil arching occurs: starting from the sides an arch-like area of loosening develops (Fig. 4.16d). Note, that this zone is not a closed arch.

The settlements above the crown increase constantly during contraction (Fig. 4.20, A/1-A/4), but the incremental settlement troughs become narrower. Hence, the incremental increase of the volume of settlement trough becomes smaller, too (Fig. 4.15).

4.3.1.2 Dilation (tail gap grouting)

At the beginning of dilation (i.e. tail gap grouting) the incremental displacements are localised in the vicinity of the tunnel (Fig. 4.16c). As a result, the surface settlements do not change (Fig. 4.20) and the volume of the settlement trough remains constant (Fig. 4.15). The zone where displacements occur is of ellipsoidal form.

The arch that formed at contraction is completely reversed during dilation (see Fig. 4.16d). The dilation of the cylinder results in a densification of the sand, which is especially pronounced laterally to the tunnel (Fig. 4.14b and Figs. 4.19a to 4.19d). The densification influences the surface settlements. Due to the densification settlements due to contraction are only slightly reversed by dilation (tail gap grouting).

The total displacements show the soil movement during the contraction-dilation cycle (Fig. 4.17a). While the soil at the surface moved toward the axis, a movement to the side can be perceived close to the crown (Fig. 4.17b).

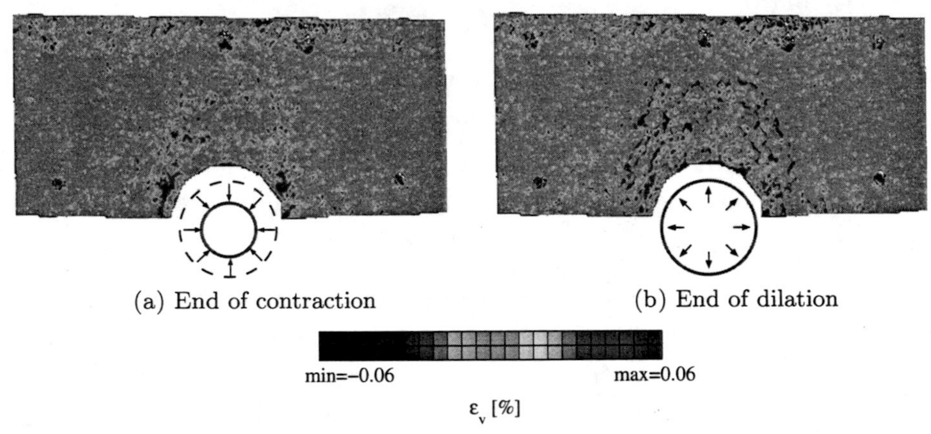

Fig. 4.14: Total volumetric strain of test **OR066-3**: (a) At the end of contraction an arch starts forming at the side. (b) The arch is reversed and the sides are densified after dilation.

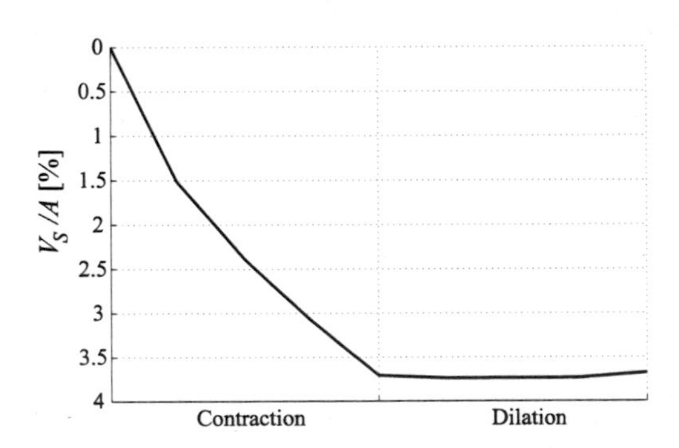

Fig. 4.15: Volume of settlement trough V_S vs. tube dilation/contraction (test **OR066-3**): The incremental increase of volume of the settlement trough is getting smaller at the end of contraction due to arching. During dilation (tail gap grouting) the volume of the settlement trough remains constant and is not reduced.

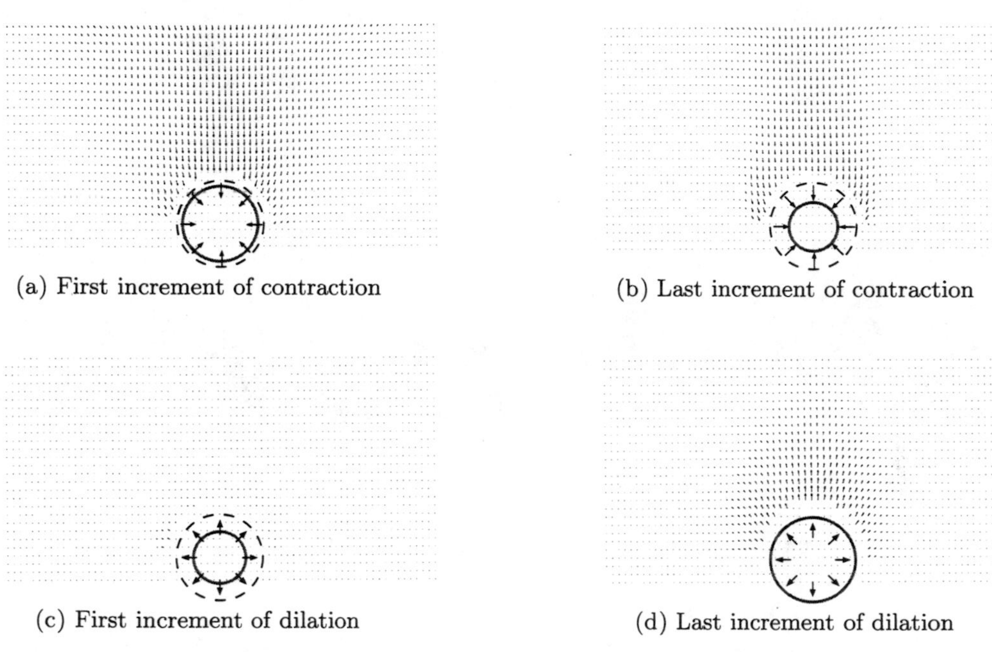

(a) First increment of contraction

(b) Last increment of contraction

(c) First increment of dilation

(d) Last increment of dilation

Fig. 4.16: Incremental displacements of test **OR066-3**: (a) Movements occur in a wide trapezoidal zone at the beginning of contraction. (b) At the end of contraction soil movement is concentrated above the cylinder. (c) The soil in the vicinity of the cylinder gets densified, (d) but the upward movement does not reach the surface.

(a) End of contraction

(b) End of dilation

Fig. 4.17: Total Displacements of test **OR066-3**: (a) The movement is concentrated above the cylinder. (b) Dilation (tail gap grouting) does not annihilate total soil movements.

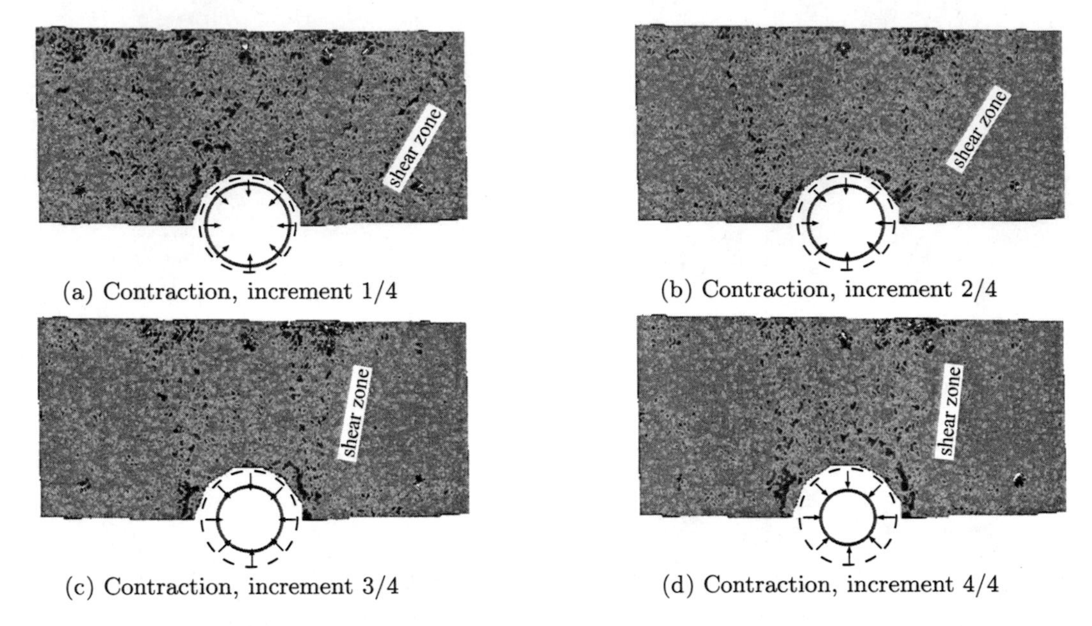

(a) Contraction, increment 1/4
(b) Contraction, increment 2/4
(c) Contraction, increment 3/4
(d) Contraction, increment 4/4

Fig. 4.18: Incremental volumetric strain $\varepsilon_{v,\mathrm{inc}}$ of test **OR066-3**: The incremental shear zones become steeper from increment to increment (a-d). Finally an arch starts forming at the sides of the cylinder (d).

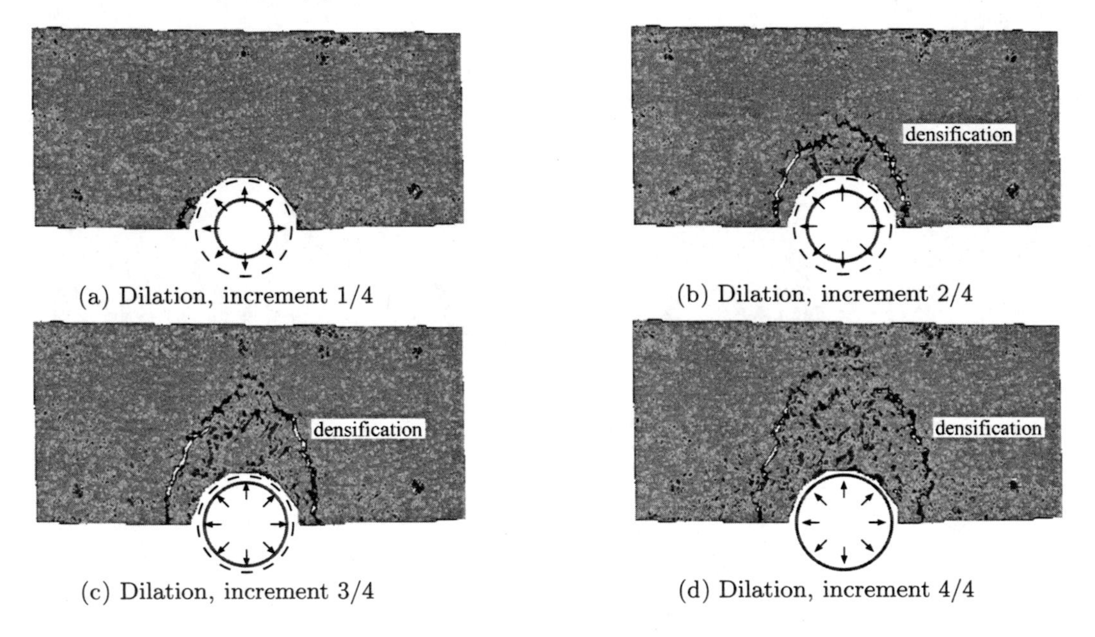

(a) Dilation, increment 1/4
(b) Dilation, increment 2/4
(c) Dilation, increment 3/4
(d) Dilation, increment 4/4

Fig. 4.19: Incremental volumetric strain $\varepsilon_{v,\mathrm{inc}}$ of test **OR066-3**: The densification moves upwards like a wave (a-d), but does not reach the surface (d).

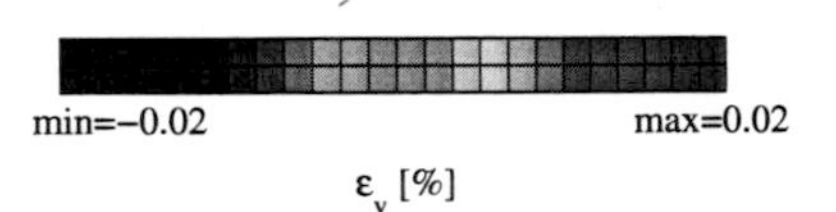

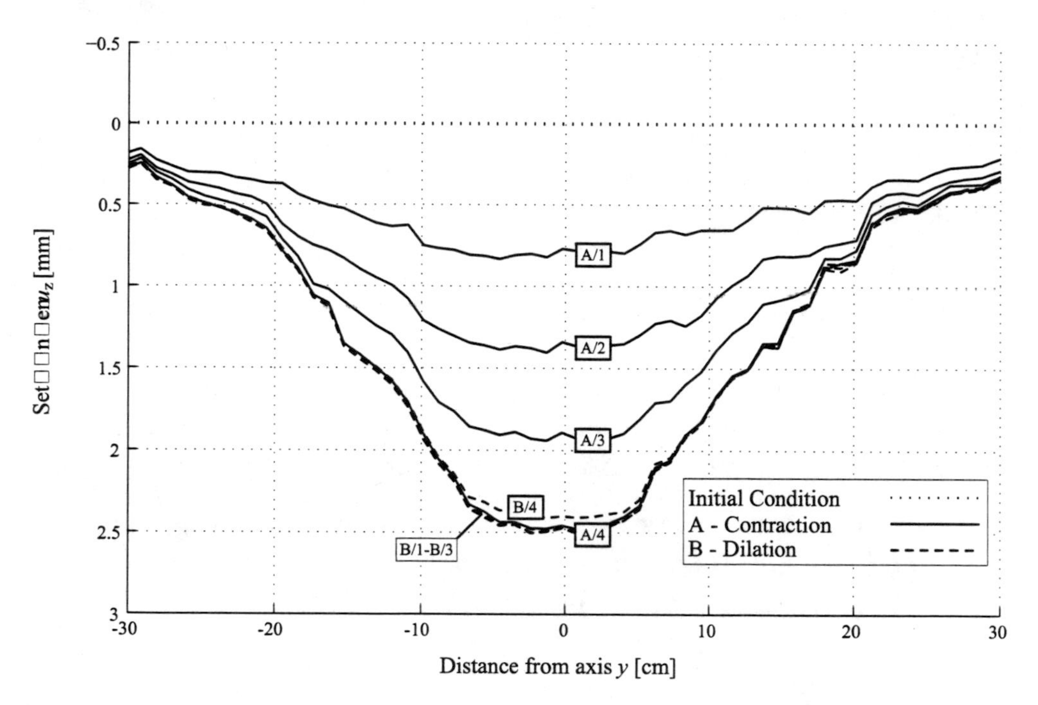

Fig. 4.20: Settlements of test **OR066-3**: The settlements increase during contraction (A/1 - A/4) and remain constant during dilation (tail gap grouting: B/1 - B/4).

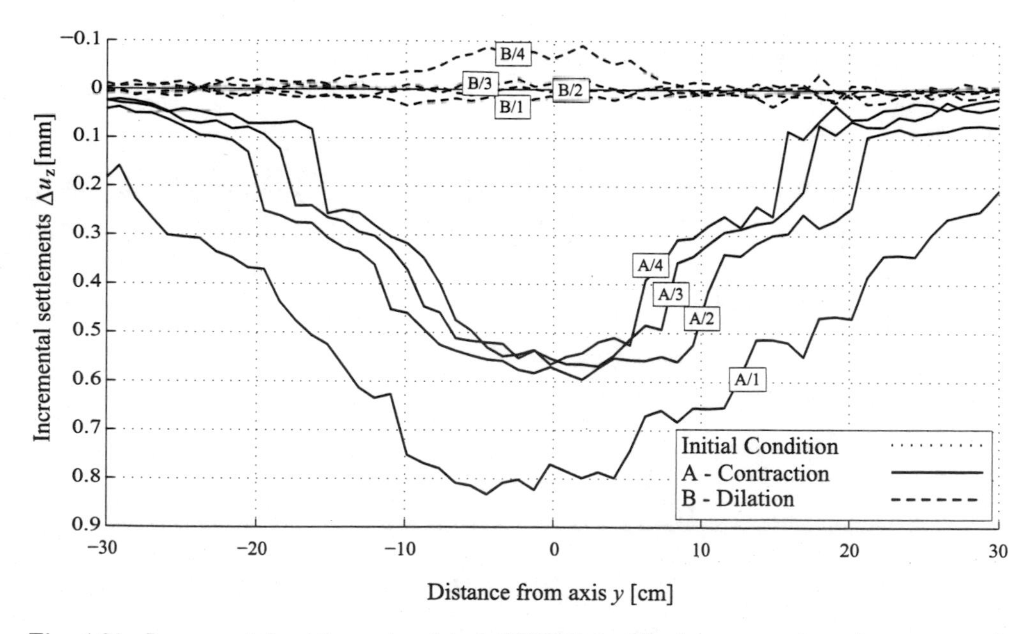

Fig. 4.21: Incremental settlements of test **OR066-3**: The incremantal settlement trough becomes narrower with ongoing contraction (A/1 - A/4). The reduction of the settlements due to dilation (tail gap grouting) are negligible (B/1 - B/4).

4.3.2 “Dense” initial state

Test **OT066-2** has been chosen to represent the tests with “dense” initial state. In this test a diameter change of 0.66 cm was applied. Ottendorf Okrilla sand was used with a density index $I_e = 0.75$.

4.3.2.1 Contraction

The soil movement is concentrated in a shaft above the tunnel from the beginning of contraction (Figs. 4.24a and 4.24b). The movements are concentrated in the vicinity of the tunnel and the surface is nearly unaffected (Fig. 4.28).

The incremental volumetric strains show an arch-shaped loosening zone above the crown (see Fig. 4.26), which, contrary to the loose tests, has a closed form. This indicates the formation of an arch above this loosening zone that carries the weight of the overlaying soil and hinders large settlements. Further smaller arches may be perceived above this arch.

4.3.2.2 Dilation (tail gap grouting)

At dilation the soil movements due to contraction are reversed and less lateral displacements than in test **OR066-3** occur. The total displacements remaining after dilation indicate a downward shift of the tunnel axis during the contraction-dilation cycle (Fig. 4.25b).

The incremental volumetric strains show the formation of a failure mechanism at the end of dilation. The block (Fig. 4.27d) is wider than the previously formed shaft and leads to a wide and pronounced upheaval at the end of dilation (Fig. 4.29). Therefore the surface area outside the shaft is upheaved at the end, while the surface above the tunnel reaches the original position again. This upheaval annihilates the volume of settlement trough from contraction (Fig. 4.22).

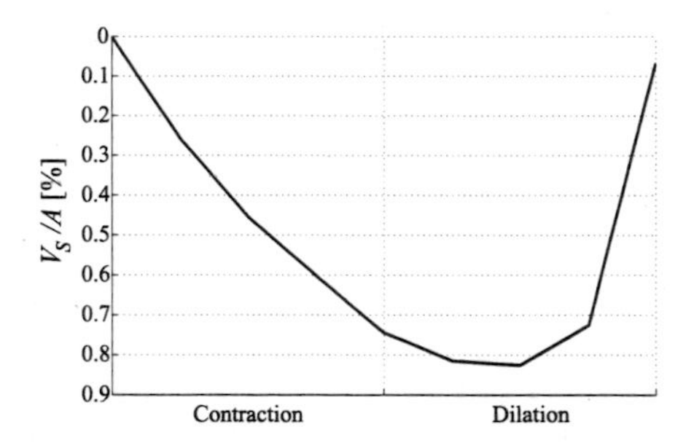

Fig. 4.22: Volume loss of test **OT066-2**: In the first contraction the volume of the settlement trough increases. At the beginning of dilation it remains constant but it is reversed at the end of dilation due to the block movement.

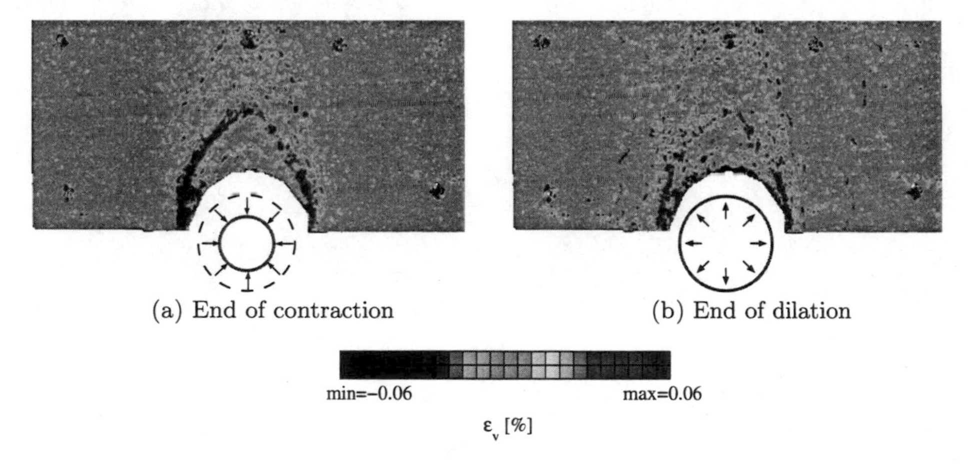

Fig. 4.23: Total volumetric strain of test **OT066-2**: (a) After contraction an arch shaped loosening zone above the cylinder occurs, (b) which is not reversed by dilation.

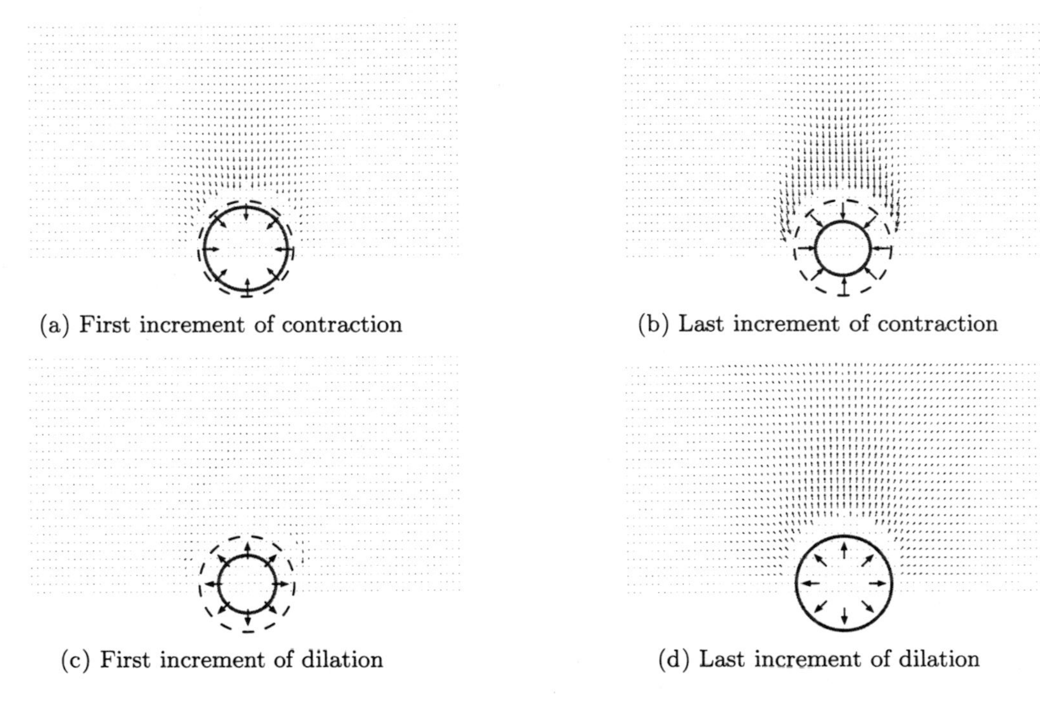

(a) First increment of contraction

(b) Last increment of contraction

(c) First increment of dilation

(d) Last increment of dilation

Fig. 4.24: Incremental Displacements of test **OT066-2**: The incremental displacements are concentrated above the cylinder (a,b). At the end of dilation a widespread upward movement indicates the formation of a block (d) that is lifted up.

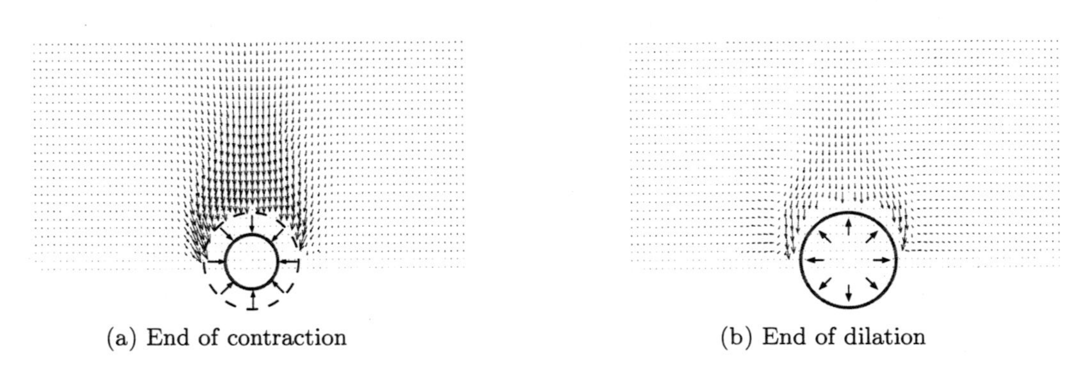

(a) End of contraction

(b) End of dilation

Fig. 4.25: Total Displacements of test **OT066-2**: (a) The displacements are concentrated above the cylinder at the end of contraction. (b) The displacements at the crown at the end of dilation indicate a downward movement of the cylinder axis. Due to the block movement the surface settlements are successfully remedied despite this axis shift.

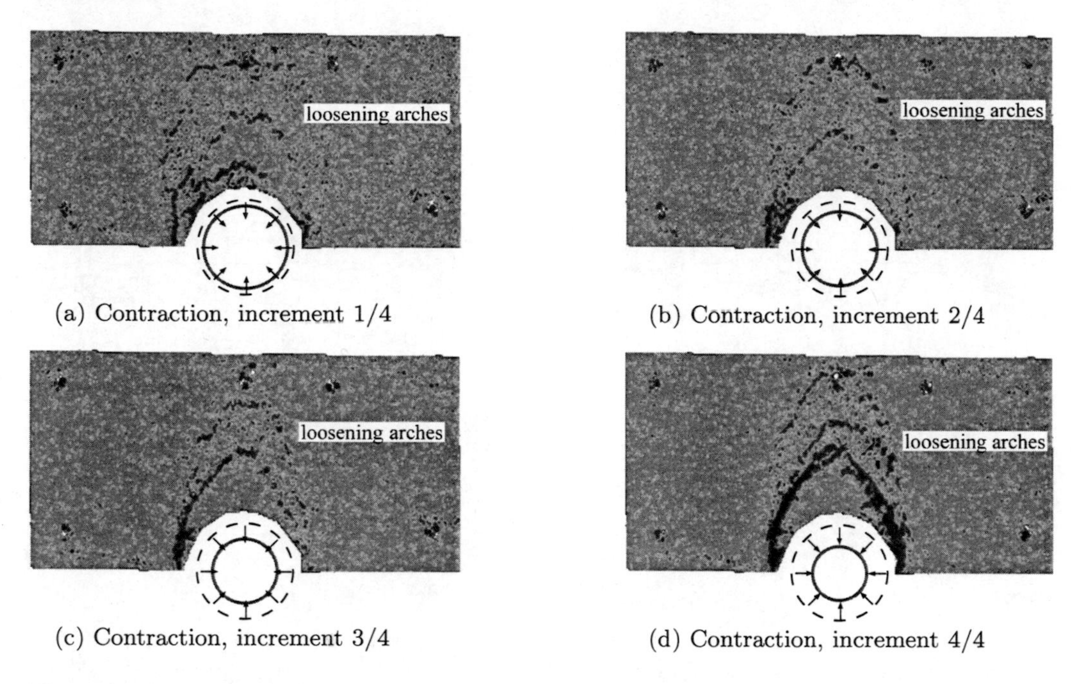

(a) Contraction, increment 1/4

(b) Contraction, increment 2/4

(c) Contraction, increment 3/4

(d) Contraction, increment 4/4

Fig. 4.26: Incremental volumetric strain $\varepsilon_{v,\mathrm{inc}}$ of test **OT066-2**: Several arches develop during contraction of the cylinder (a-c). Finally loosening is concentrated in one arch (d).

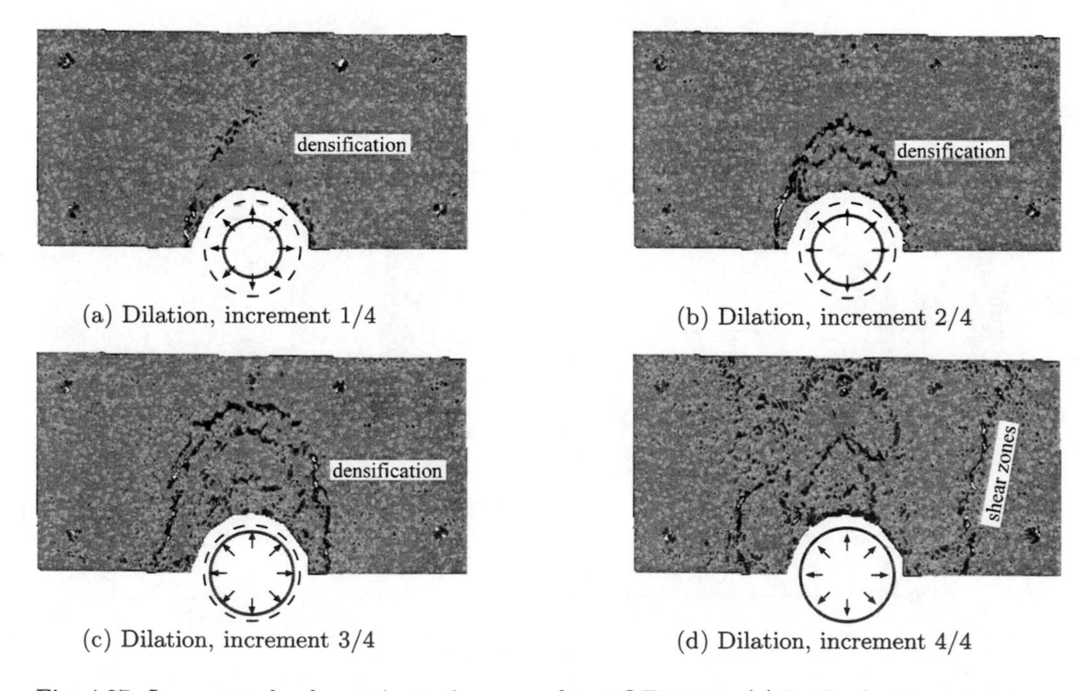

(a) Dilation, increment 1/4

(b) Dilation, increment 2/4

(c) Dilation, increment 3/4

(d) Dilation, increment 4/4

Fig. 4.27: Incremental volumetric strain $\varepsilon_{v,\mathrm{inc}}$ of test **OT066-2**: (a) In the first increment the dilation leads to a densification of the dominant arch. (b-c) Then the surrounding soil of the cylinder is densified. (d) Finally a block is lifted by the dilation.

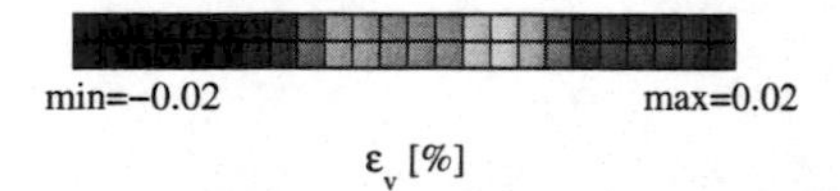

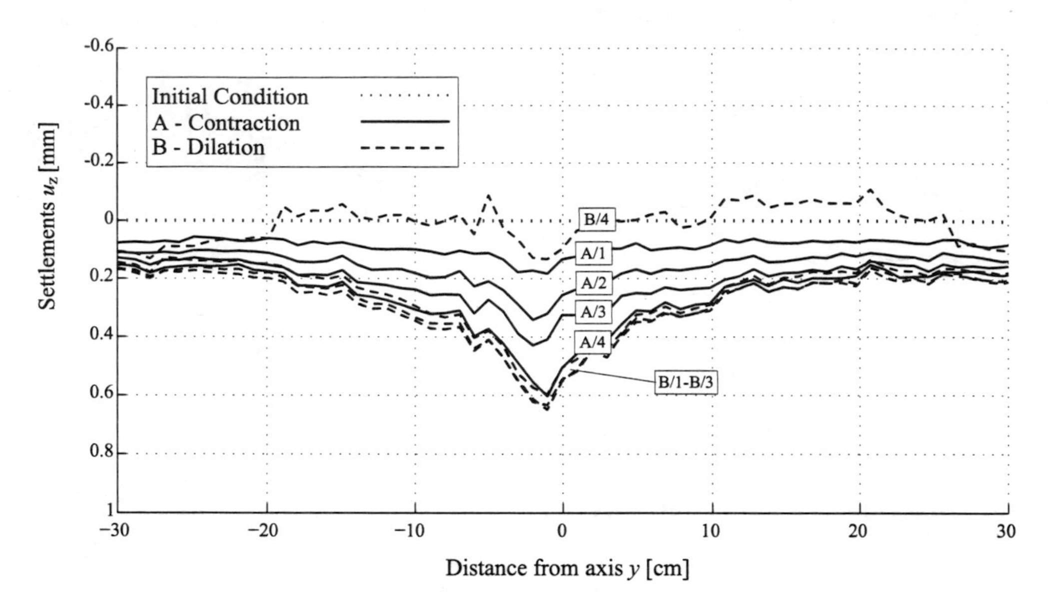

Fig. 4.28: Settlements of test **OT066-2**: The settlements increase during contraction of the cylinder (A/1 - A/4). At the beginning of dilation they remain constant (B/1 - B/3), but at the end of dilation an abrupt upheaval occurs (B/4).

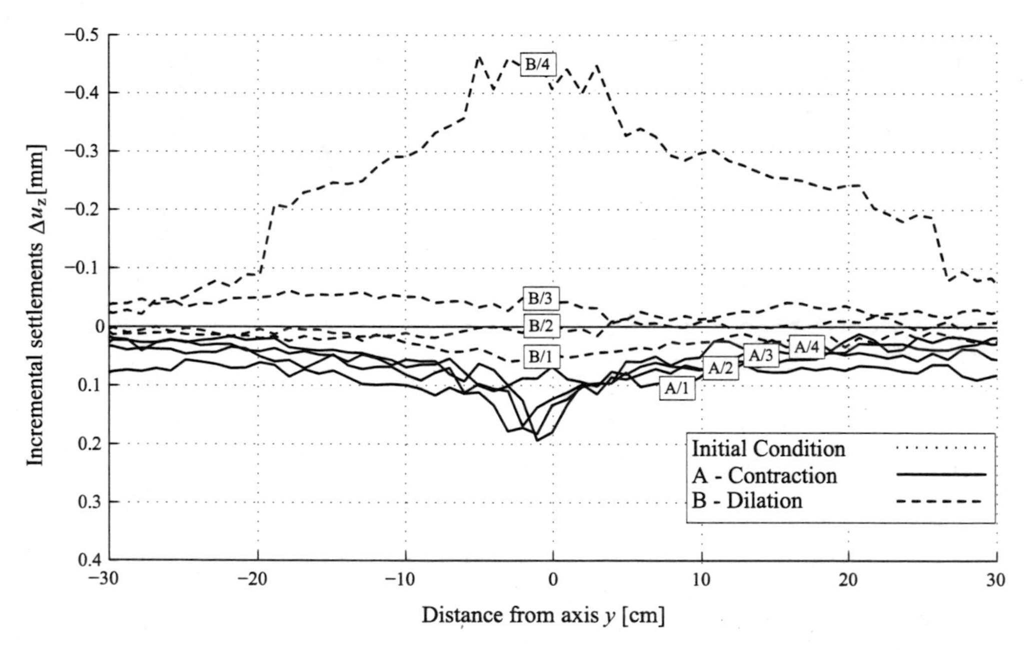

Fig. 4.29: Incremental settlements of test **OT066-2**: The incremental settlements are nearly constant during contraction (A/1 - A/4). At the beginning of dilation the settlements still increase (B1), whereas at the end of dilation an upheaval is shown (B4).

4.3.3 Summary of the test results

As the tests are carried out with different changes of diameter, the results cannot be directly compared. A widespread technique to get comparable results is to make them dimensionless. For this purpose, the variables ΔD and ΔA are introduced (see Fig. 4.30). ΔD is the change of diameter and ΔA is the change of the cross-section of the tunnel. Now the settlements can be normalised with ΔD and the volume of settlement trough with ΔA.

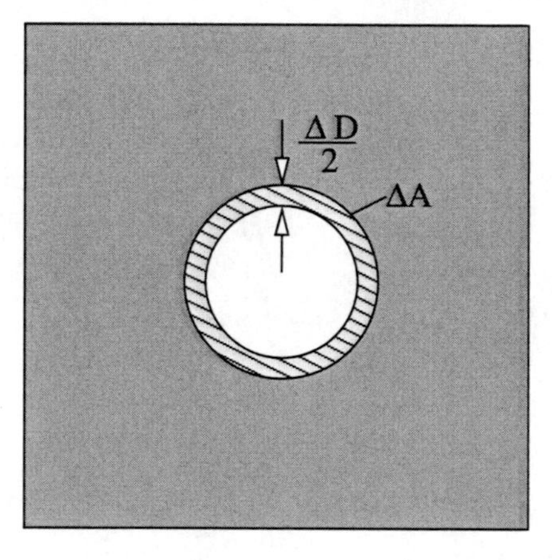

Fig. 4.30: Variables used for normalisation.

Fig. (4.31) and Fig. (4.32) show plots of relative density index vs. normalised volume of settlement trough V_S and normalised settlements, respectively. The tests show the dependence of the volume V_S of settlement trough on the relative density. The linear regression lines, which were calculated using the method of least squares, point out that volume of settlement trough V_S and settlement are slightly reduced by dilation of the cylinder. Furthermore the regression lines show, that the higher the relative density index is the more volume of settlement trough V_S decreases by dilation. At very high values of I_e one even gets volume win.

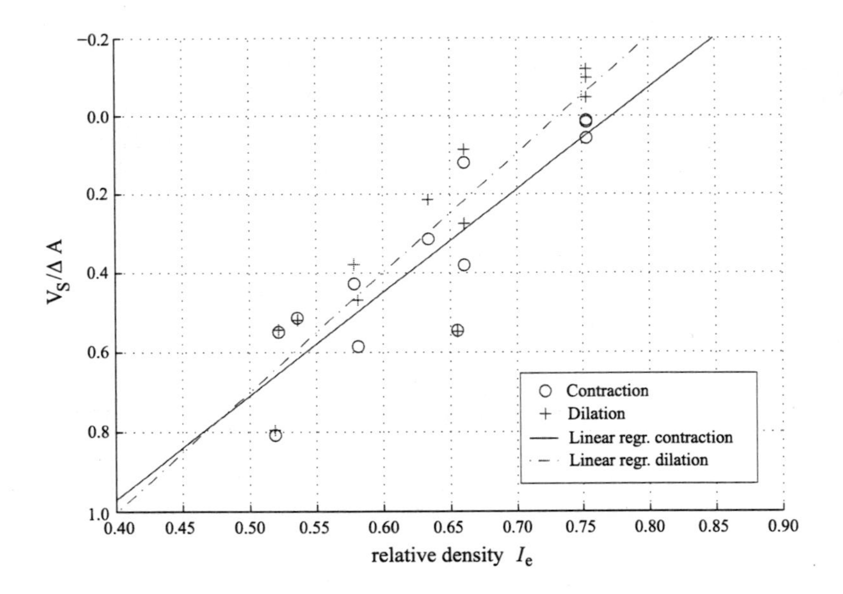

Fig. 4.31: Volume of settlement trough V_S vs. relative density index I_e.

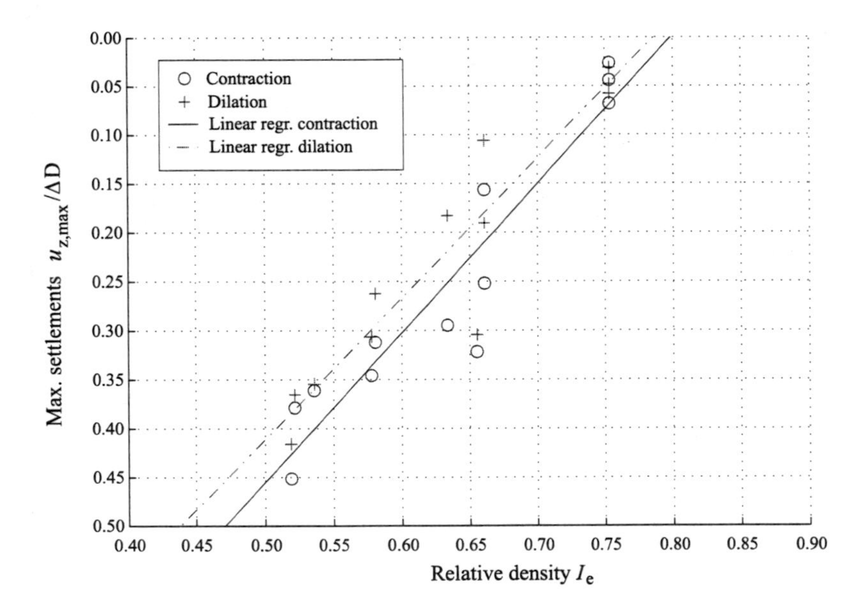

Fig. 4.32: Max. settlements u_z vs. relative density index I_e.

4.3.4 Influence of diameter change

To study the influence of contraction on the settlements, three test series with similar initial densities are compared: **OR0.5**, **OR0.66** and **OT1.0** (cf. Tab. 4.2). Remember that **O** stands for Ottendorf-Okrilla sand, **R** stands for build in with a rake, **T** stands for tamping and the number stands for the change of diameter in centimetres. The initial unit weights of these tests range from 16.7 kN/m^3 to 17.2 kN/m^3 (see Fig. 4.13).

Fig. (4.33a) and Fig. (4.33b) show that settlements and volume of settlement trough do not increase linearly with the diameter change ΔD. With increasing diameter change both the volume of settlement curve and the settlement curve flatten due to the developing arch.

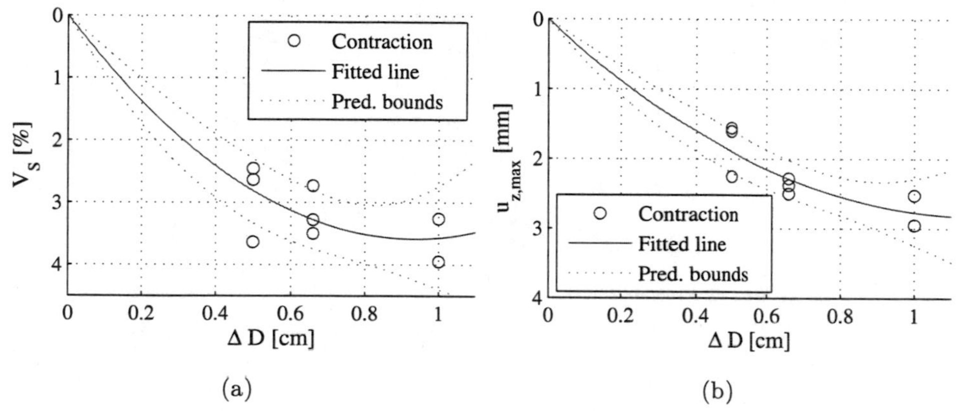

Fig. 4.33: (a) Volume of settlement trough V_S and (b) maximum settlements $u_{z,max}$ of tests with similar initial density vs. diameter change.

4.4 Conclusions and outlook

From the laboratory tests the following conclusions can be drawn:

- In the case of loose sand at the beginning of contraction a wide trapezoidal zone of soil movement develops, which is bounded by shear planes (Fig. 4.34a). The zone becomes narrower with ongoing contraction (Fig. 4.34b-d). It can be assumed, that the angle of the shear planes depends on the relative density and the ratio of horizontal to vertical stresses. With ongoing contraction the vertical stresses above the crown decrease and the horizontal stresse increase, thus the ratio of horizontal to vertical stresses increases and the shear planes become steeper. In the last increment, the shear planes are vertical and delimit a shaft that starts from the sides of the tunnel (Fig. 4.34d). Within this shaft arching sets on above the crown.

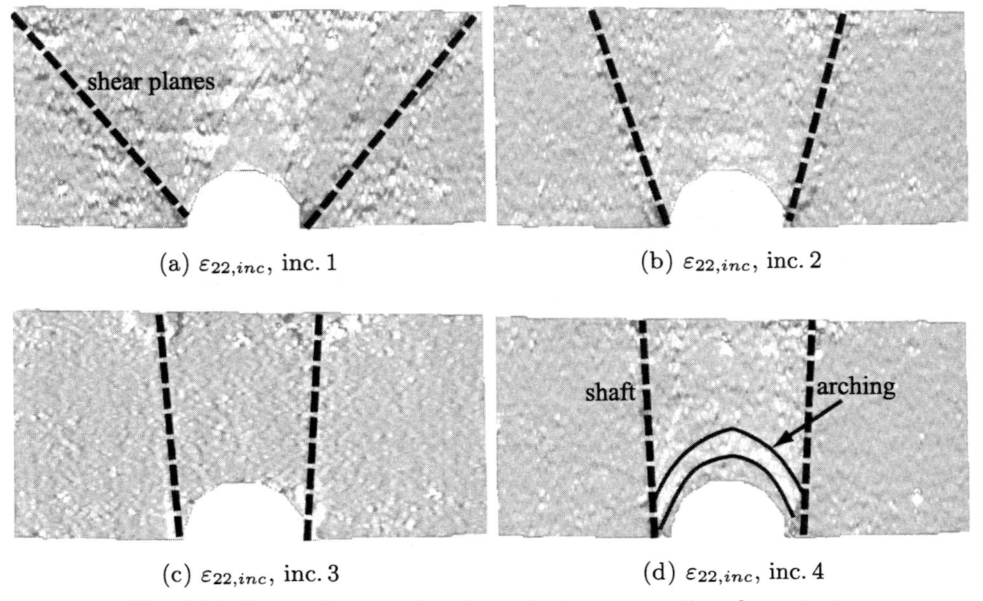

(a) $\varepsilon_{22,inc}$, inc. 1 (b) $\varepsilon_{22,inc}$, inc. 2

(c) $\varepsilon_{22,inc}$, inc. 3 (d) $\varepsilon_{22,inc}$, inc. 4

Fig. 4.34: Shear planes and arching during contraction, loose tests.

- Due to densification the settlements can not be reversed by dilation of the cylinder to its original size in a loose assembly. Especially the soil in the sides is densified.

- In the case of dense sand arching sets on with contraction immediately above the tunnel (Fig. 4.35a). The arches are limited by a shaft that starts from the sides of the tunnel and goes vertically up to the surface. Surface settlements remain very small compared to tests with loose sand. Their extension is limited to the area inside the shaft. At the end of dilation a trapezoidal quasi-rigid block can develop, which may cause upheavals (Fig. 4.35b). As this block is wider than the shaft, upheavals develop also outside the projection of the tunnel to the surface.

- In a dense sand settlements can be reversed by dilation of the cylinder to its original size, by heaving the soil as a rigid block.

- Soil deformation is governed by strain localisation that appears as shear bands or arch shaped loosening zones. Thus, the shape of the surface settlement troughs depends on the failure mechanism which develop (cf. Fig. 4.28 and Fig. 4.20).

- Due to arching the relation between change of diameter and settlements is non linear.

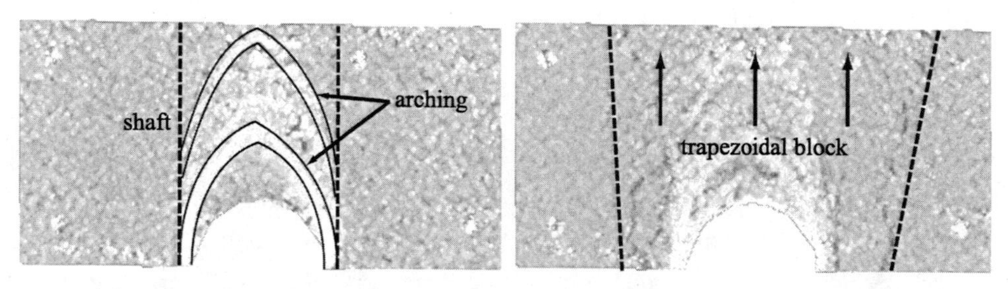

(a) $\varepsilon_{v,inc}$, inc. 1 of contraction (b) $\varepsilon_{v,inc}$, Inc. 4 of dilation

Fig. 4.35: Shear planes, arching and block movement, dense tests.

Chapter 5

Numerical simulation of model tests

A FE model of the test apparatus has been created. A comparison of qualitative and quantitative test results with the FE calculations shows the appropriateness of the chosen boundary conditions and the material model. Then parametric studies are carried out. A parametric study with a prototype initial stress state and varying densities investigates the appropriateness of the test results for site conditions. Finally, a parametric study of the earth pressure coefficient K is carried out to investigate its influence on the settlement.

5.1 FE model

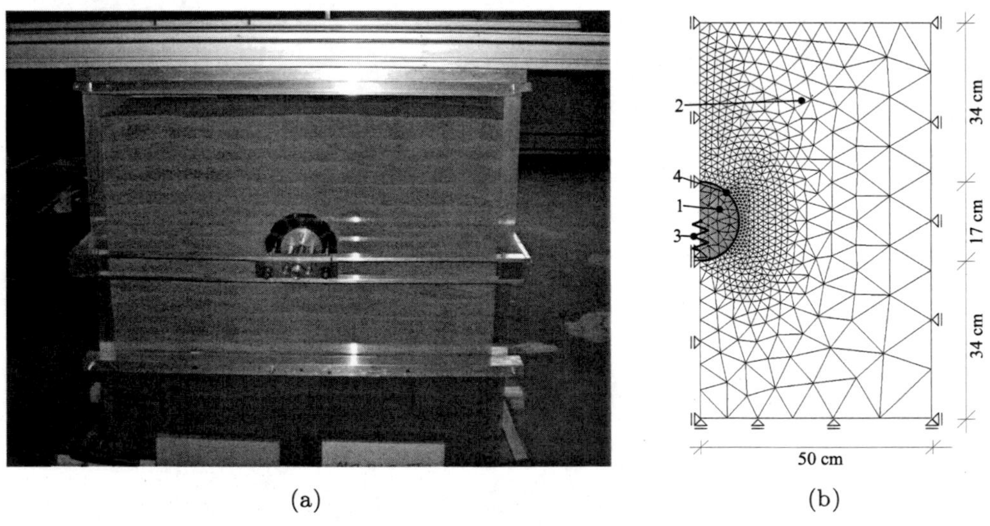

(a) (b)

Fig. 5.1: (a) Test apparatus and (b) corresponding FE-model: (1) cylinder, (2) soil, (3) spring, (4) interfaces.

The 2D FE-mesh used for this problem is shown in Fig. (5.1b). Because of symmetry, only one half of the apparatus is modelled. The FE-model of the model consists of two parts: the cylinder (tunnel) and the sand body. The cylinder was simulated by means of 52 CPE3[1] elements. Elastic material model was used with Young's modulus $E = 10^{12}$ kN/m^2. This value was chosen to keep the stress-induced deformations of the cylinder negligible. By changing the temperature, the radius of the cylinder can be varied due to thermal expansion. Thus, by choosing consistent thermal expansion coefficient and temperature change, the radial dilation/contraction of the cylinder may be numerically simulated.

The cylinder is mounted on a spring (see Fig. 5.1b,(3)). The stiffness of the spring c is set equivalent to the stiffness of the four aluminium beams of the laboratory model, which carry the cylinder, and is given by

$$c = 4 \cdot \frac{48 E_b I_b}{l_b} \cdot 0.4 \text{ m} = 81 \text{ kN/ m/ m} \quad . \tag{5.1}$$

The soil body consists of 1160 CPE3[1]elements. Hypoplasticity with intergranular strain was used to model the sand behaviour. The material parameters (Tab. 5.1) are taken from Laudahn [58], the parameters for the intergranular strain are taken from Niemunis and Herle [73]. The intergranular strains are set to zero at the beginning of the analyses.

	ϕ_c [°]	h_s [MPa]	n	e_{d0}	e_{c0}	e_{i0}	α	β
O.-O. sand	32	1000	0.19	0.42	0.75	0.86	0.19	2.25

R	m_R	m_T	β_r	χ
$1 \cdot 10^{-4}$	5.0	2.0	0.50	6.0

Tab. 5.1: Hypoplastic and intergranular strain parameters of Ottendorf-Okrilla sand ([58] and [73]).

Contact of soil and cylinder is taken into account using the surface interaction technique [see 42, chap. 21.3]. Tangential interaction is modelled by the basic COULOMB friction model, that defines a critical shear stress $\tau_{\text{crit}} = \mu \cdot \sigma_{\text{surf}}$, with σ_{surf} being the contact pressure. The contact surfaces can carry shear stresses up to τ_{crit} before they start sliding relative to each other. Normal interaction is modelled using the "softened" contact relationship [42] with

[1]CPE3: triangular constant plane strain elements.

exponential law (see Fig. 5.2). In this relationship the surfaces begin to transmit contact pressure once the clearance between them reduces to c_0. The contact pressure transmitted between the surfaces increases exponentially as the clearance diminishes. This relationship was chosen because it is able to simulate the influence of the rubber membrane applied upon the cylinder and it stabilises the FE-calculations because of its smooth transition from "no contact" to "full contact".

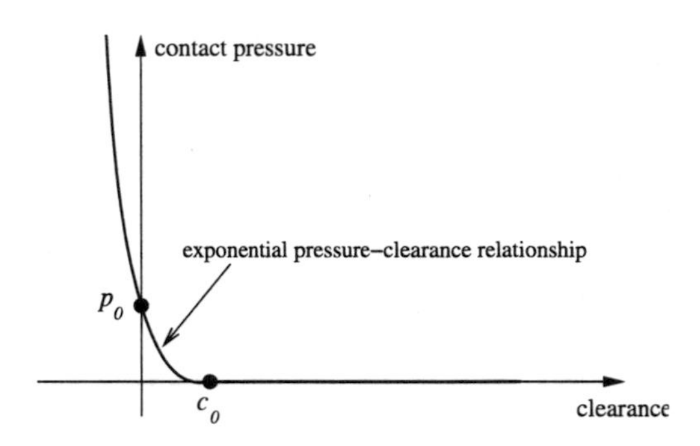

Fig. 5.2: Pressure clearance relationship of "softened" contact relationship

The following calculation steps are performed (cf. Fig. 5.3):

Initial step: In the initial step gravity is applied to both parts, the soil and the cylinder. As the resulting stresses are previously defined in the input file (initial conditions) the load is applied at once and has not to be splitted in increments. Equilibrium is immediately found and the resulting displacements are zero. Thus, the mesh is still undeformed. To circumvent iterations at the contact surfaces, the nodes of the contact surfaces are fixed (Fig. 5.3).

Release step: In this step the boundary conditions applied to the cylinder surface are released and surface interaction is enabled.

Contraction step: By applying a temperature reduction the cylinder is contracted.

Dilation step: By applying a temperature increase the cylinder is dilated.

In total 14 analyses have been carried out. All analyses have been performed with a change of diameter of 0.66 cm.

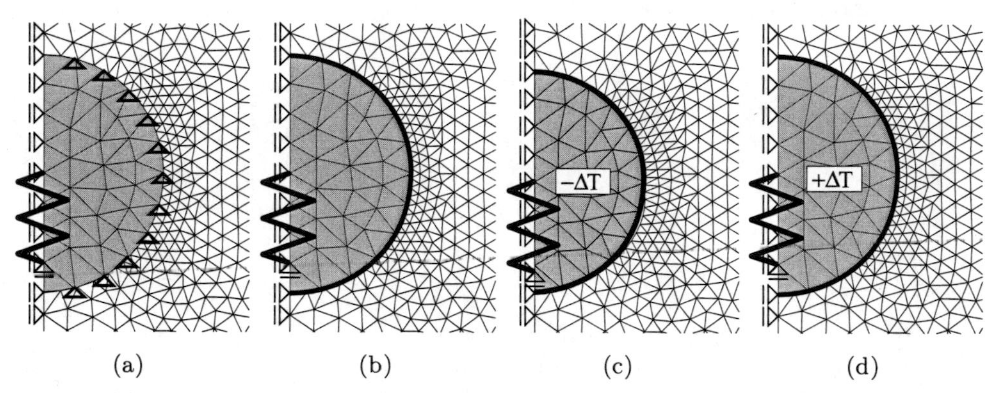

Fig. 5.3: (a) Surface nodes are fixed at initial step. ; (b) In the release step surface nodes are released and surface interaction algorithm is activated. (c) Contraction is simulated by applying a temperature reduction to the cylinder. (d) Dilation is simulated by applying a temperature increase.

First the appropriateness of the model is shown by a parametric study with laboratory conditions. That is, the initial stresses are calculated from the soil density and gravitation $1g$. K was set to $K = 1 - \sin\varphi = 0.46$ for all calculations and the relative initial density index I_e is varied in the range from 0 to 1 in the following steps $I_e = [0.03; 0.2; 0.4; 0.55; 0.75; 0.97]$. Then a parametric study with prototype stress level has been performed. The gravitation applied in these analyses is $50g$, thus the stress state corresponds to a tunnel of 8.50 m diameter having a tail gap of 12.5 cm, 16.5 cm or 25 cm, respectively, and 17 m overburden. The earth pressure coefficient K was set to $K = 1 - \sin\varphi = 0.46$ and the relative initial densities are set to $r_e = [0.4; 0.55; 0.75; 0.97]$. Note, that due to the high stress level the stress-dependent density index r_e is used instead of the relative density index I_e. Finally a parametric study with varying the earth pressure coefficient K has been carried out. The following values of K have been used: $K = [0.46; 0.75; 1.0; 2.0]$. A gravity of $50g$ has been applied and the initial stress dependent density index was set to $r_e = 0.6$ for all analyses.

	Density	Gravitation	Pressure coeff. K
Laboratory	$I_e = [0.03; 0.2; 0.4; \ldots$ $\ldots 0.55; 0.75; 0.97]$	$1g$	0.46
Site	$r_e = [0.4; 0.55; 0.75; 0.97]$	$50g$	0.46
K **study**	$r_e = 0.6$	$50g$	[0.46;0.75;1.0;2.0]

Tab. 5.2: Overview on analyses carried out with FE model.

5.2 Results

5.2.1 "Loose" initial state

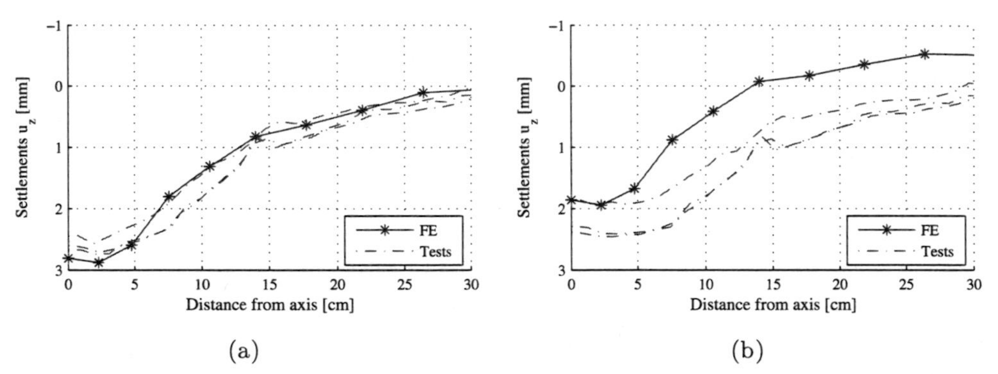

Fig. 5.4: Settlement troughs from FE analysis with $I_e = 0.55$ and test results of tests with "loose" initial state (test series **OR0.66**): (a) Settlement after contraction of cylinder, (b) settlement after dilation of cylinder

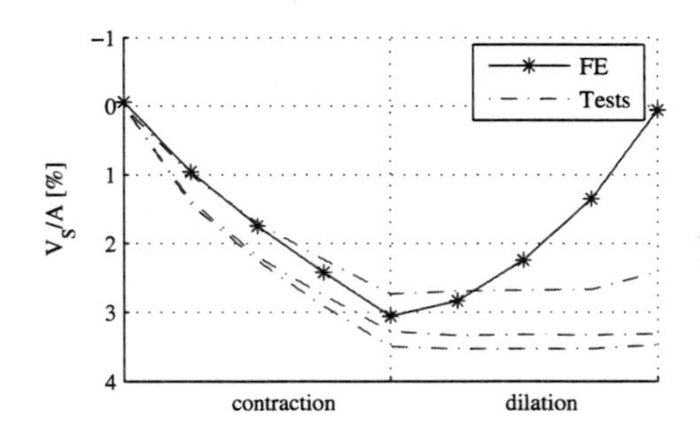

Fig. 5.5: Volume of settlement trough from FE analysis $I_e = 0.55$ and test series **OR0.66**.

The qualitative behaviour of soil in the FE model is examined by a comparison of FE analyses and laboratory tests with similar initial densities. The initial density indices I_e of the model test series **OR0.66**[2] are in the range from 0.52 to 0.57. Their qualitative soil behaviour corresponds to the class of tests with "loose" initial state (section 4.3.1). The tests are compared to a FE analysis with an initial relative density index of $I_e = 0.55$.

[2]Ottendorf-Okrilla sand; built in with a rake; diameter change of 0.66 cm.

The settlement trough of the FE analysis after contraction shows good agreement with the test results, and so does the graph of the volume of the settlement trough (Fig. 5.4a and Fig. 5.5). Only at the end of dilation the reduction of the settlement trough is overestimated (Fig. 5.4b).

The vertical strain ε_{zz} after contraction is plotted in Fig. (5.6b). This plot matches the test results fairly good (Fig. 5.6a). The volumetric strain after dilation (Fig. 5.7b) show some discrepancy to the test results (Fig. 5.7a). Contrary to the model tests no densified areas (shown by the black areas in Fig. 5.7a) can be found in the results of the FE analysis. This behaviour may result from the low stress level. Note that the material parameters were calibrated for a stress range from 500 - 1500 kPa [58].

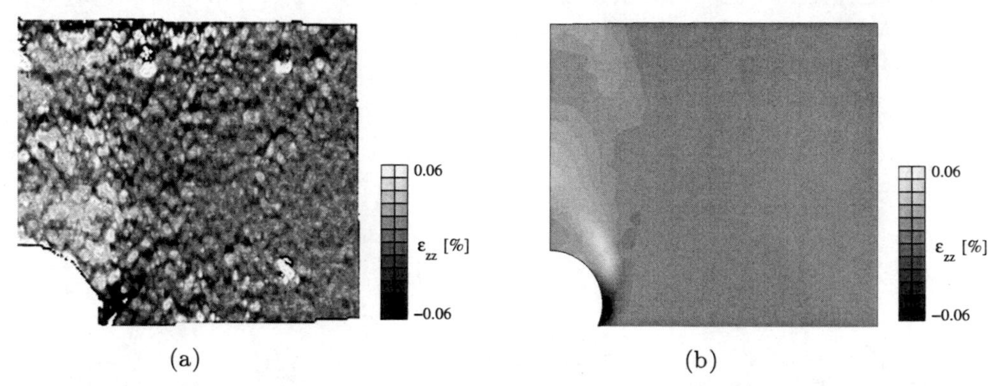

Fig. 5.6: ε_{zz} after contraction: (a) Test OR066-3, (b) FE analysis, $I_e = 0.55$.

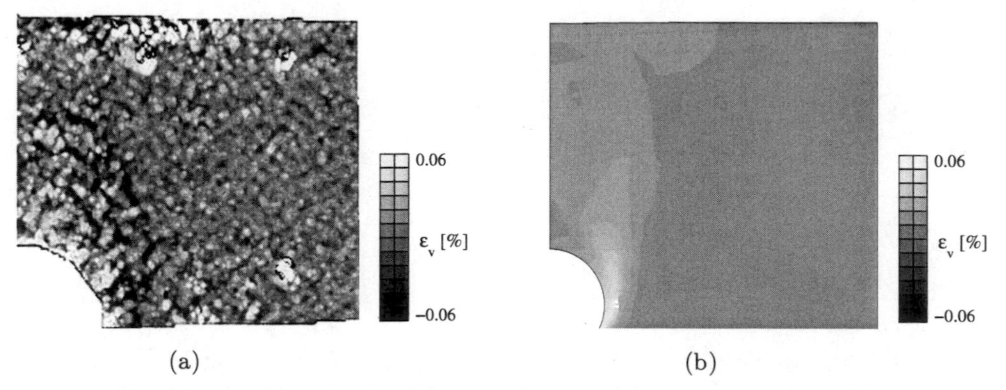

Fig. 5.7: ε_v after dilation: (a) Test OR066-3, (b) FE analysis, $I_e = 0.55$.

5.2.2 “Dense” initial state

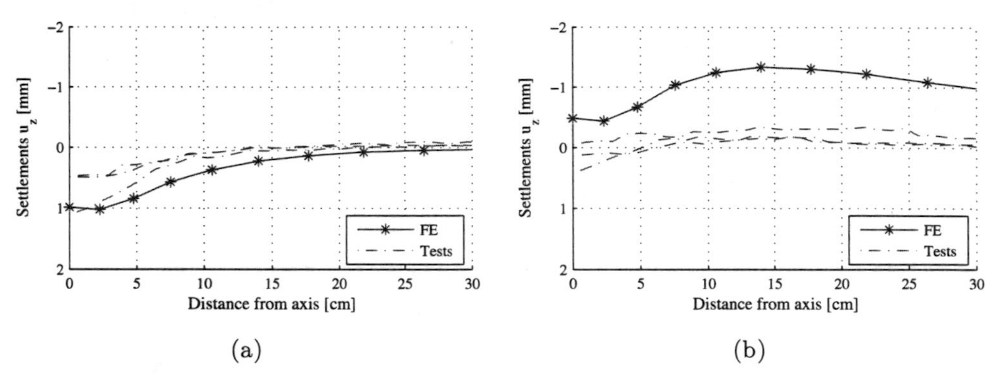

Fig. 5.8: Settlement troughs from FE analysis with $I_e = 0.75$ and test results of tests with “dense” initial state (test series **OT0.66**): (a) Settlement after contraction of cylinder, (b) settlement after dilation of cylinder

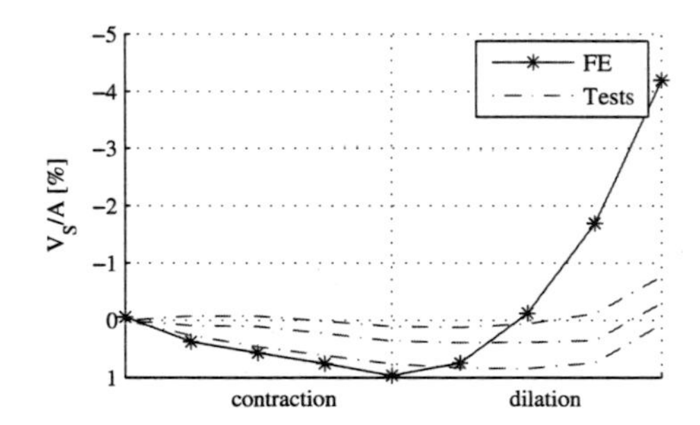

Fig. 5.9: Volume V_S of settlement trough from FE analysis with $I_e = 0.75$ and test series **OT0.66**.

The model test series **OT0.66** may be considered as “dense” initial state tests (section 4.3.2). All tests of this series were built in with an initial relative density index of 0.75. Their results are compared to a FE analysis with an initial relative density index of $I_e = 0.75$.

Again, the settlement trough of the FE analysis shows good agreement with the test results after contraction (Fig. 5.8a). But the upheaval at the end of dilation (Fig. 5.8b) is overestimated and thus the resulting volume of settlement trough is not realistic (Fig. 5.9).

The volumetric strain of FE analysis and model test after contraction is plotted in Fig. (5.10). The arching effect observed in the test is reproduced by the FE-calculation, even the size of the arch equals the one from the model test.

The white spots in Fig. (5.11a) border a block that is moving upwards at the end of dilation. The origin of this block formation is also shown in the FE-analysis (Fig. 5.11b).

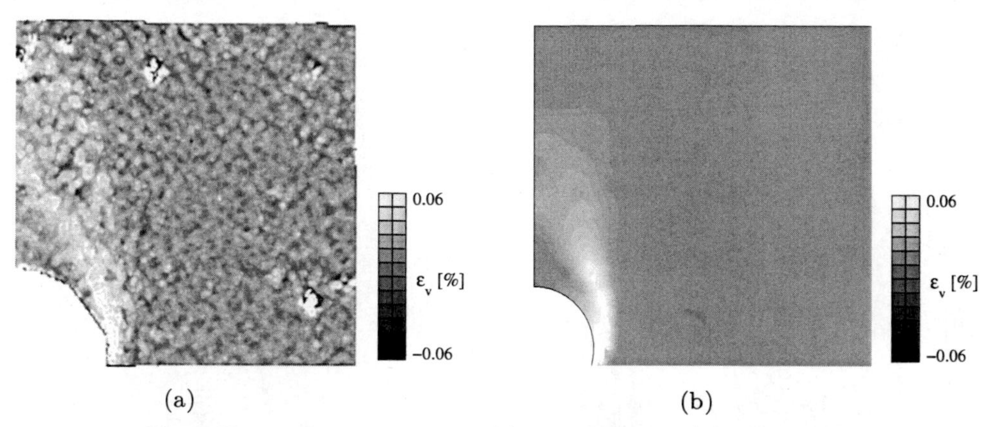

(a) (b)

Fig. 5.10: ε_v after contraction: (a) Test OT066-2, (b) FE analysis

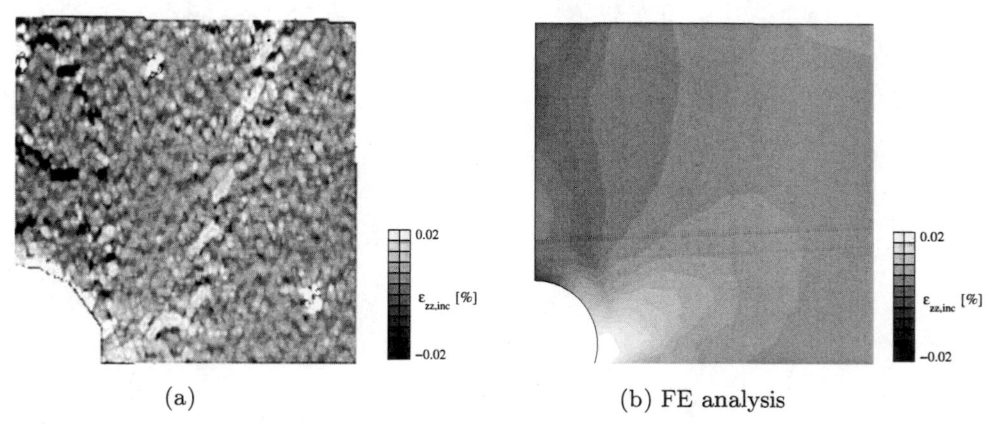

(a) (b) FE analysis

Fig. 5.11: $\varepsilon_{zz,\mathrm{inc}}$ after dilation: (a) Test OT066-2, (b) FE analysis

5.2.3 Summary of FE-calculations with laboratory conditions

Beneath the above described FE analysis, further calculations with varying densities have been carried out. The calculated maximum settlements and the volume V_S of the settlement trough after contraction show very good agreement to the ones from the laboratory tests (Figs. 5.12a and 5.12b). Remember that the material parameters are not fitted, they are taken from Laudahn [58].

While the maximum settlements after dilation show quite a good agreement to the ones from the tests (Fig. 5.13a), the heave due to tunnel dilation is overestimated and thus the volume V_S of settlement trough is underestimated by the FE analyses (Fig. 5.13b). As discussed above, this may result from the low stress level and the fact that the material parameters are not fitted to this conditions.

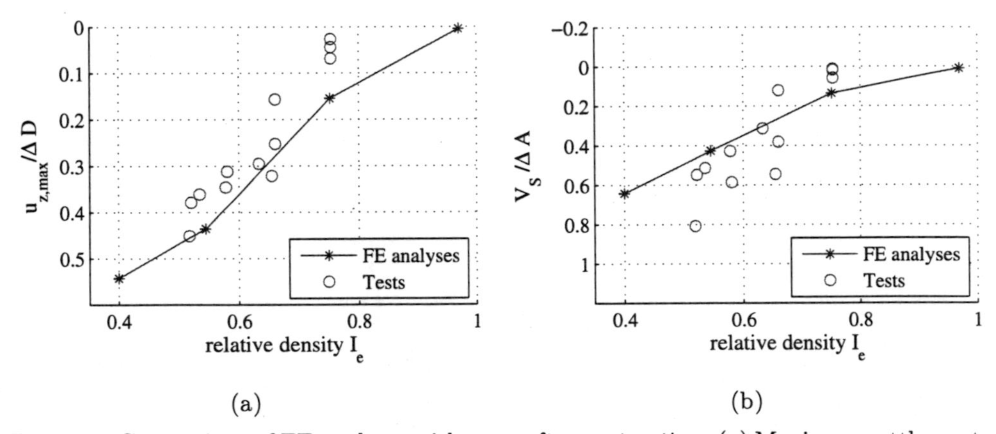

Fig. 5.12: Comparison of FE analyses with tests *after contraction*: (a) Maximum settlements and (b) Volume V_S of settlement trough.

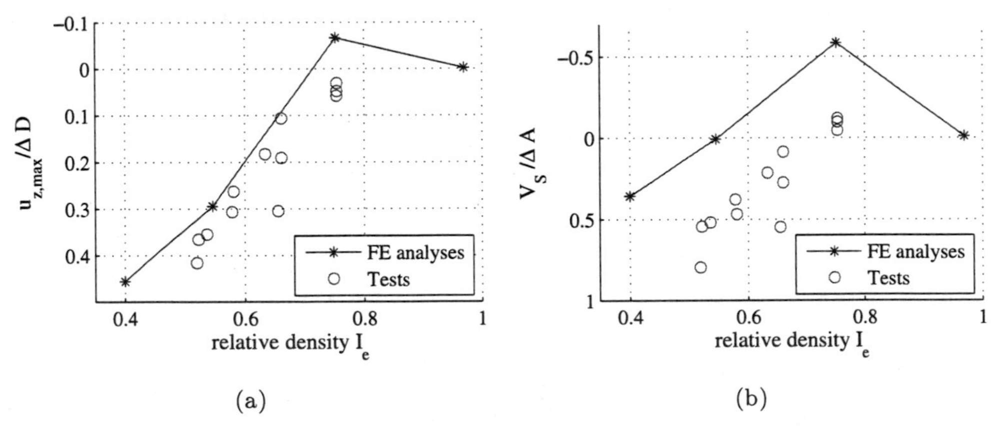

Fig. 5.13: Comparison of FE analyses with tests *after dilation*: (a) Maximum settlements and (b) Volume V_S of settlement trough.

5.2.4 Influence of stress level

All laboratory tests and FE analyses have been carried out under normal gravitational conditions, so called $1g$ tests. Due to stress dependence (barotropy), these tests and analyses have limited applicability. Thus, FE analyses with prototype stresses have been carried out. The earth pressure coefficient for the initial state has been set to $K = 0.46$. In this parametric study the stress dependent density has been used to define the initial density, which has been varied as follows: $r_e = [0.4; 0.55; 0.75; 0.97]$.

Due to higher stresses the soil behaves softer and thus the settlements from prototype analyses ($50g$) are larger than the ones from the model analyses ($1g$) (Fig. 5.14a). Moreover the upheaval of the surface due to dilation of the cylinder is smaller in the prototype analyses than in the model analyses. This results from the lower angle of dilatancy at higher stresses. Thus, "dense" soil behaviour (section 4.3.2) is not likely to occur at tunnel sites. One may conclude that non-cohesive soils behave at stress levels of shallow tunnel sites like the model tests with "loose" initial state.

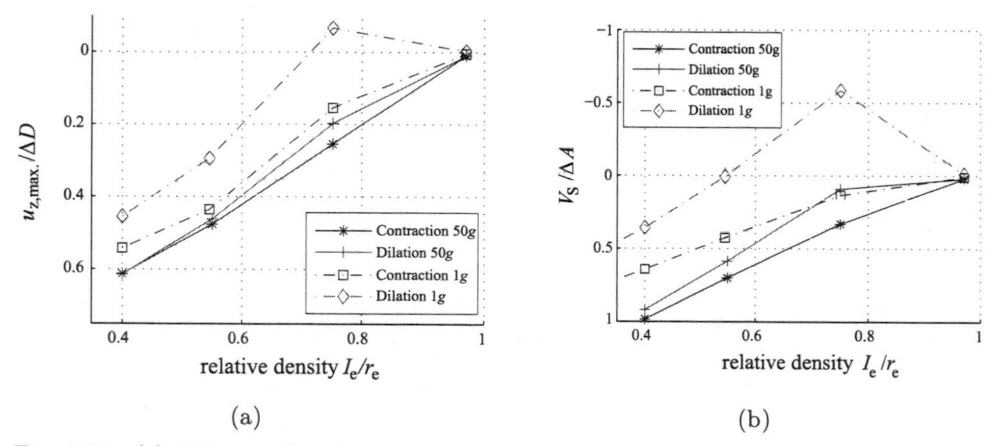

Fig. 5.14: (a) Volume of settlement trough V_S vs. relative density I_e ($1g$ analyses) and relative stress dependent density r_e ($50g$ analyses); (b) Settlement above crown vs. relative density I_e ($1g$ analyses) and relative stress dependent density r_e ($50g$ analyses).

5.2.5 Influence of earth pressure coefficient K

The influence of the earth pressure coefficient K is studied by a number of analyses with various initial lateral earth pressure coefficients. Figs. (5.15a) and (5.15b) show that form and extent of the settlement trough depends on K. The higher the lateral pressure is, the smaller are the settlements after the contraction (Fig.5.16a). Due to the lateral restrain the settlement trough from analyses with high K values are wider than the ones with low K values. As a result the volumes V_S of the settlement trough after contraction from high K are nearly the same, as from low K (Fig.5.16b).

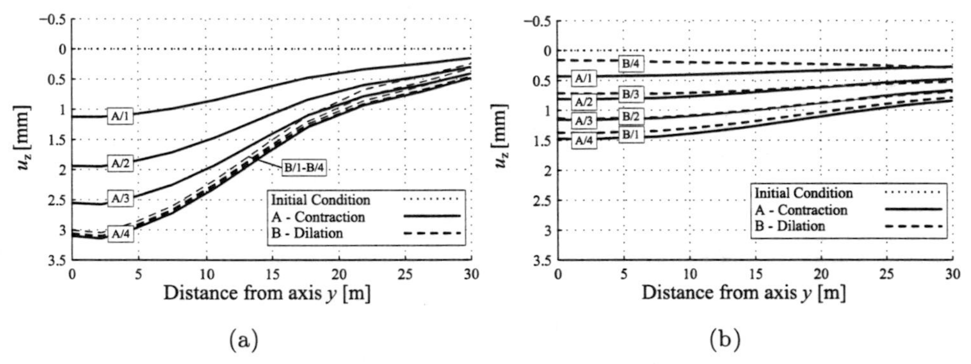

Fig. 5.15: Settlement troughs for (a) $K = 0.46$ and (b) $K = 2.0$.

Furthermore the earth pressure coefficient K has considerable influence on the soil behaviour at dilation of the cylinder (i.e. tail gap grouting). While the surface is more or less unaffected at low earth pressure $K = 0.46$ (Fig. 5.15a), the settlements are nearly reversed at high lateral earth pressure coefficient $K = 2.0$ (Fig. 5.15b).

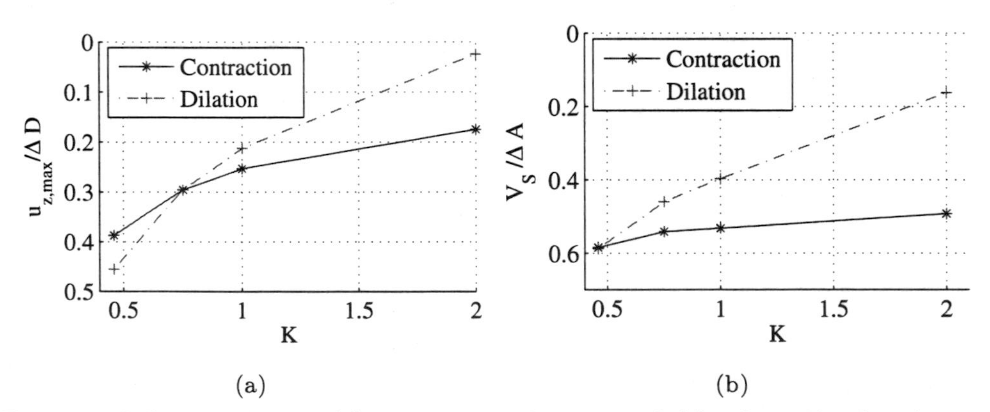

Fig. 5.16: Influence of K on (a) maximum settlements and (b) volume V_S of settlement trough.

5.3 Conclusions

The FE analyses of this chapter showed, that

- Hypoplasticity is an appropriate tool to model cyclic dilation/contraction of a cylinder and thus to model shield tunnelling,
- non-cohesive soils behave at stress levels of shallow tunnel sites like the model tests with "loose" initial state. Hence, tail gap grouting is not successful for such soils.
- the earth pressure coefficient K influences magnitude and extend of the settlement curve. The higher K, the wider is the settlement trough and the smaller are the maximum settlements. Thus, a higher K value means a flatter settlement curve and minor danger for adjacent buildings.

Chapter 6

Pre-arching, a concept to reduce settlements

The previous chapters showed that tail gap grouting is not always successful in limiting or reducing settlements. The success of its application depends on the geotechnical site conditions, especially on soil density and stress state.

In many cases further measures are necessary to ensure serviceability and stability of adjacent buildings. Practicable geotechnical measures are improvement of the endangered building or ground treatment measures. The latter include all methods in which the ground response is controlled by means of modifying the soil characteristics (e.g. grouting or freezing). A widely used method is compensation grouting.

Compensation grouting is a technique for controlling ground movements. A liquid grout is pumped into the ground through narrow tubes (Tubes à Manchette or TAMs) between the advancing tunnel and the ground surface. In general the tubes are arranged in a horizontal layer (see Fig. 6.1a). At tunnel excavation movements occur due to various mechanisms of ground loss. The intention of grouting is to prevent any significant movements from propagating to the surface, thus preventing damage to overlying structures. A detailed monitoring of surface movements in conjunction with careful control of grout injection is required as the ground movements are not reduced but compensated. Unfortunately compensation grouting using a horizontal layout of TAMs causes an increase of the hoop lining stresses [55]. In NATM tunnelling this may even cause a collapse of the lining.

In this chapter an alternative arrangement of the TAMs is proposed and investigated. Own model tests and FE analysis showed the influence of soil arching on the settlements. The faster a soil arch develops, the smaller are the settlements. However, soil movement — which we want to avoid — is necessary for the formation of an arch. The basic idea of the alternative method termed pre-arching is to generate a soil arch prior to the construction of the tunnel by grouting. The grouting zones are chosen perpendicular to the arch line (see Fig. 6.1b). Due to the grouting the soil becomes pre-stressed and less soil movement is necessary for the formation of an arch.

FE analyses show the effect of pre-arching on the settlement and on the hoop stresses of tunnel lining.

6.1 Boundary conditions

The mesh in Fig. (6.2) is used to simulate pre-arching. Soil is modelled by means of 1838 CPE3 elements [1]. Hypoplasticity with intergranular strain is used to model soil behaviour. The soil parameters of Ottendorf-Okrilla (Tab. 5.1, p. 95) sand are chosen for the analyses.

So-called hydrostatic fluid elements are used to simulate grouting. Hydrostatic fluid elements are elements on the boundary of a fluid-filled cavity (Fig. 6.3b). These elements provide the coupling between the deformation of the fluid-filled cavity and the pressure exerted by the contained fluid on the boundary of the cavity. The fluid inside the cavity can be compressible or incompressible, with the fluid volume given as a function of the fluid pressure, the fluid temperature and the fluid mass in the cavity. In these analyses a bulk modulus of $2.2 \cdot 10^6$ kN/m^2 was chosen, which equals the bulk modulus of water[2]. After grouting the fluid-filled elements are replaced by CPE3 elements with linear elastic material model to simulate the hardened grout. Young's modulus of the hardened grout is chosen according to Falk [27], who reports the stiffness of grouted sand to take a value of two to four times the stiffness of original soil. From numerical element tests the stiffness of Ottendorf-Okrilla sand in the investigated stress level is approximately 50000 kN/m^2, thus a stiffness of 150000 kN/m^2 is chosen for the hardened grout.

The lining is modelled by means of 20 B21[3] beam elements. The lining is

[1] CPE3 ... 3-node constant plane strain element

[2] The bulk modulus of water was chosen, because no value for the bulk modulus of fresh grout could by found by the author. However, in this boundary value problem the bulk modulus is of minor importance, as the injected volume is fitted in a way that surface settlements are minimised. A change of the bulk modulus only results in a change of the injected grout volume. If the method is used for design purpose, the bulk modulus of the fresh grout should be determined by means o laboratory tests.

[3] B21 ... 2-node linear beam

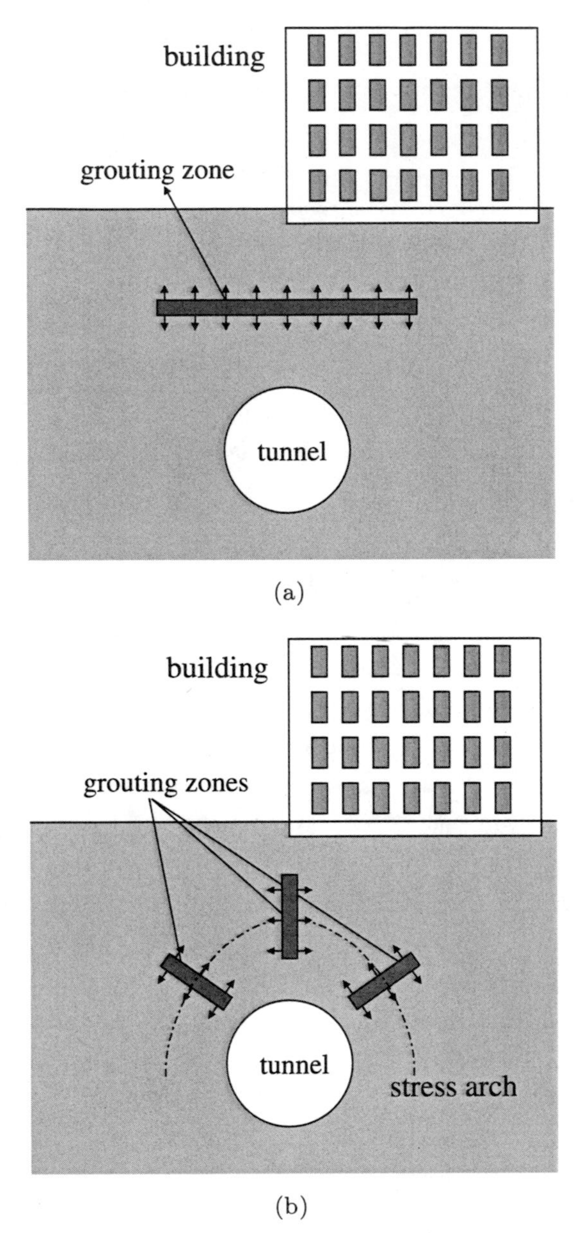

Fig. 6.1: Compensation grouting with (a) horizontal layout of TAMs and (b) layout according to pre-arching.

40 cm thick, linear elastic material model is used and the Young's modulus equals $E = 30 \cdot 10^6$ kN/m^2. Contact between lining and soil is simulated with the surface interaction technique, friction is neglected.

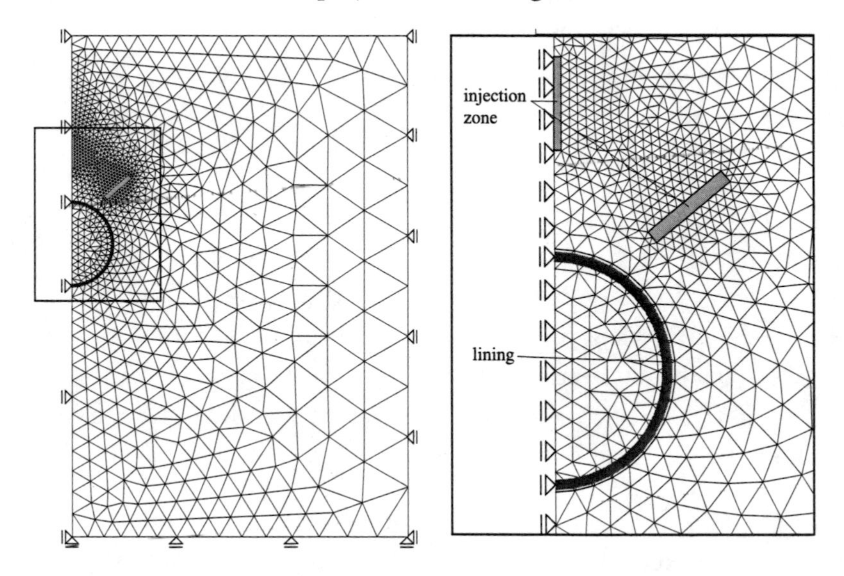

Fig. 6.2: Mesh and boundary conditions to simulate pre-arching.

Compaction grouting is modelled in 3 steps. First the soil elements of the grouted area are replaced by hydrostatic elements (Fig. 6.3b). Then the volume of the fluid cavity is increased by 0.055 m^3 (zone 1) and 0.255 m^3 (zone 2), respectively, to simulate compaction grouting (Fig. 6.3c). In a third step the hydrostatic elements are replaced by CPE3 elements (Fig. 6.3d), that simulate the hardened grout.

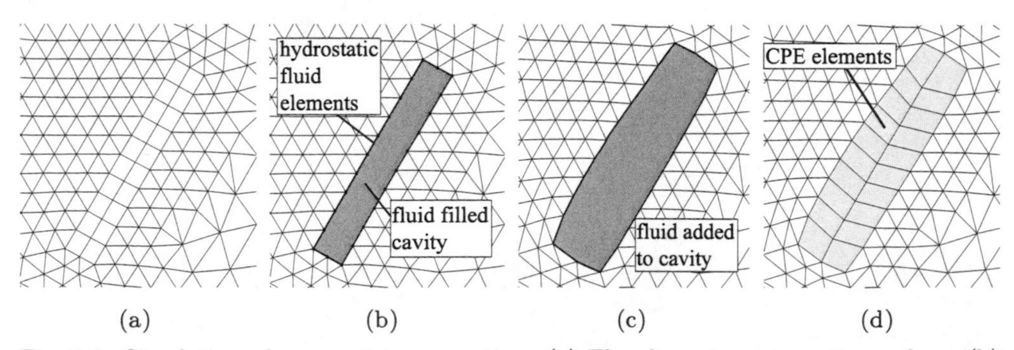

Fig. 6.3: Simulation of compensation grouting: (a) The elements representing soil are (b) replaced by a fluid-filled cavity, where (c) fluid mass is pumped in. (d) Finally the fluid-filled cavity is replaced by elements representing the hardened grout.

After finishing compaction grouting, tunnelling starts. Tunnelling is simulated in 2 steps. First the elements of the tunnel cross section are removed and the gap between soil and tunnel lining closes. The interaction between lining and soil is modelled with the surface interaction technique of ABAQUS (section 5.1, p. 95), friction is set to zero. Then the tunnel lining is shrinked. The lining is not fixed in vertical direction and thus may move upward due to buoyancy.

Due to grouting, soil displacements occur and the nodes representing the tunnel contour in the further course of the simulation are displaced as well. Thus, the excavation area is displaced prior to excavation. On the one hand this may cause numerical problems when the lining is activated, because the two surfaces (tunnel contour and lining) are overlapping. On the other hand, the results of the simulations with pre-arching cannot be compared to the ones without pre-arching any more, as different cross sections are excavated.

To overcome this problem, the FE analysis is stopped after grouting and the displacements of the nodes representing the tunnel contour $(u_y^{\text{prel.}}, u_z^{\text{prel.}})$ are read out (Fig. 6.4b). Then the displacements are subtracted from the initial coordinates of these nodes (Fig. 6.4c) and the calculation starts from the scratch with the corrected coordinates of the contour nodes. Now the nodes are approximately in the desired position after grouting and the simulation may continue with tunnel excavation (Fig. 6.4d). A summary of all conducted steps is given in Fig. (6.5).

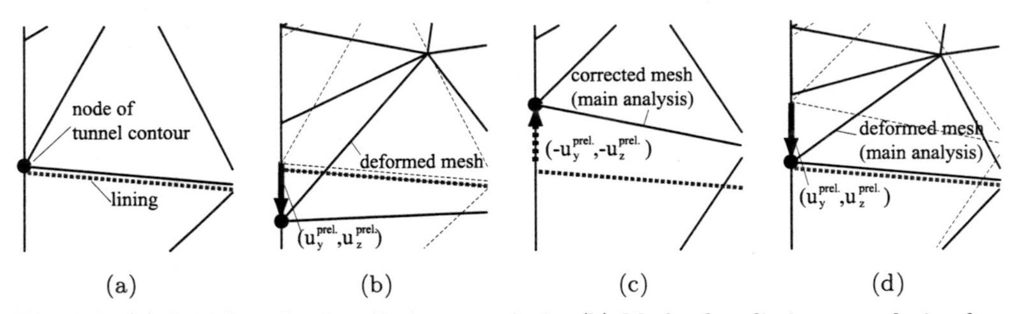

Fig. 6.4: (a) Initial mesh of preliminary analysis; (b) Mesh of preliminary analysis after compaction grouting and hardening. The tunnel contour is placed inside the lining; (c) Initial mesh of main analysis; (d) Mesh of main analysis after compaction grouting and hardening. The tunnel contour is placed in the desired position and the simulation may continue with tunnel excavation.

Preliminary analysis:

1. *Initial conditions*
2. *Preparation of grouting*
   Replacement of CPE elements in injection zone with hydrostatic fluid elements.
3. *Compaction grouting*
   Adding fluid to hydrostatic fluid elements.
4. *Hardening of grout*
   Replacement of hydrostatic fluid elements with CPE elements, $E = 1.5 \cdot 10^6$ kN/m^2.

Adjustment of mesh:
The initial postion $(y_{0,\text{cont.}}, z_{0,\text{cont.}})$ of the nodes representing the tunnel contour line is adjusted:

$$y_{0,\text{cont.}}^{\text{main}} = y_{0,\text{cont.}}^{\text{prel.}} - u_{y,\text{cont.}}^{\text{prel.}}$$

$$z_{0,\text{cont.}}^{\text{main}} = z_{0,\text{cont.}}^{\text{prel.}} - u_{z,\text{cont.}}^{\text{prel.}}$$

Main analysis:

1. *Initial conditions*
2. *Preparation of grouting*
   Replacement of CPE elements in injection zone with hydrostatic fluid elements.
3. *Compaction grouting*
   Adding fluid to hydrostatic fluid elements.
4. *Hardening of grout*
   Replacement of hydrostatic fluid elements with CPE elements that simulate hardened grout, $E = 1.5 \cdot 10^6$ kN/m^2.
5. *Activation of lining*
   Lining elements and contact algorithm (lining - tunnel contour) are activated.
6. *Excavation*
   Soil elements of tunnel crossection are removed.
7. *Shrinkage of lining*
   Radius of tunnel lining is decreased by applied temperature change.

Fig. 6.5: Process chart of FE analyses to simulate compaction grouting.

6.2 Results

The results of the analysis are compared to an analysis of tunnel excavation without pre-arching. In this analysis the same model is used, but steps 1 - 4 (Fig. 6.5) are not considered and only the tunnelling process is simulated.

The trajectories of the minimum principal stresses in Fig. (6.6a) show, that compaction grouting of the injection zones creates an arch. After tunnel excavation arching remains. The localised soil movement below the arch leads to a loosened zone just beneath the arch (Fig. 6.7b).

Compensation grouting causes an upheaval of the surface (Fig. 6.8a). The following settlements due to tunnelling are about 50 % of the settlements obtained without pre-arching (Figs. 6.8a and 6.9a). Note, that the tunnel-induced settlement trough (termed relative settlement) obtained with pre-arching is flatter but wider than the one obtained without pre-arching. Therefore the tunnel induced volumes of the settlement trough do not differ very much (Fig. 6.10b).

However, a flatter settlement trough causes less damage to adjacent buildings. The 2nd derivation of the vertical displacements is used to assess potential building damage in Fig. (6.11). The values of the analysis with pre-arching are by far smaller than the ones without pre-arching.

The maximum hoop stresses (Fig. 6.12b) and the maximum bending moments (Fig. 6.13b) in the lining in the pre-arching calculations are slightly bigger than the ones without pre-arching. This may result from the stresses induced by injection zone 1 on the lining.

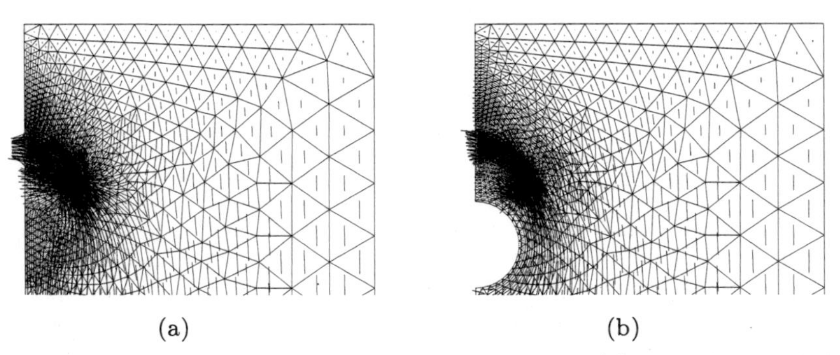

Fig. 6.6: Trajectories of minimum principal stresses after (a) pre-arching and after (b) tunnel excavation.

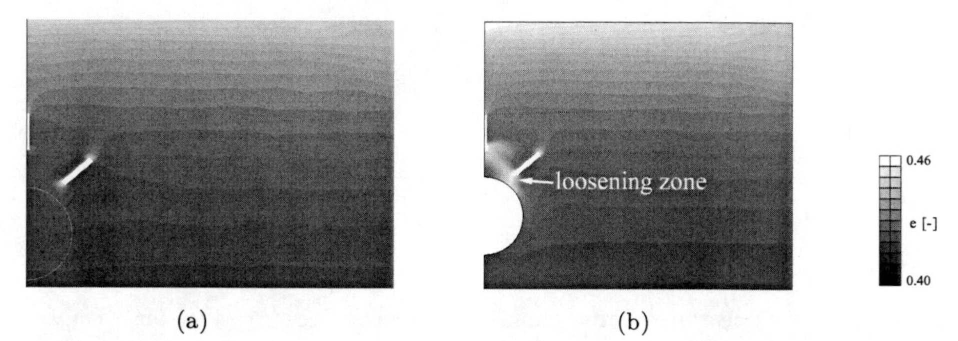

(a) (b)

Fig. 6.7: Void ratio after (a) pre-arching and after (b) tunnel excavation.

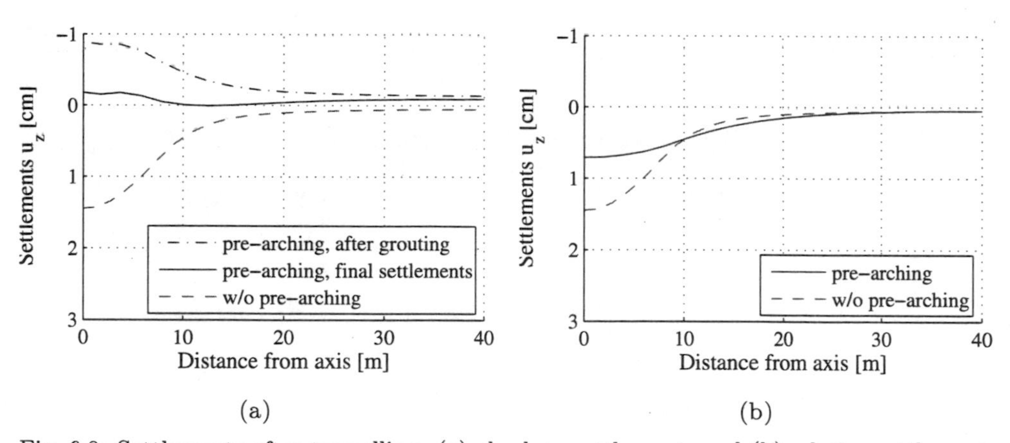

(a) (b)

Fig. 6.8: Settlements after tunnelling: (a) absolute settlements and (b) relative settlements (i.e. tunnel-induced).

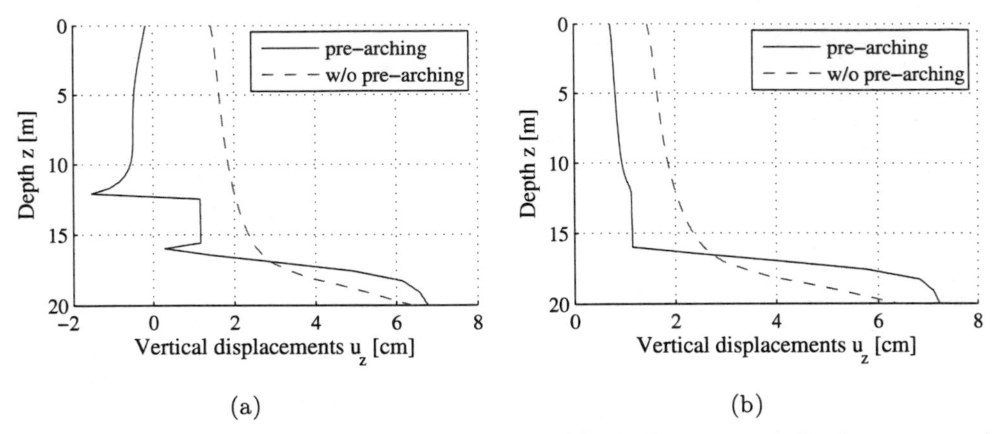

(a) (b)

Fig. 6.9: Vertical displacements after tunnelling: (a) absolute vertical displacements and (b) relative vertical displacements.

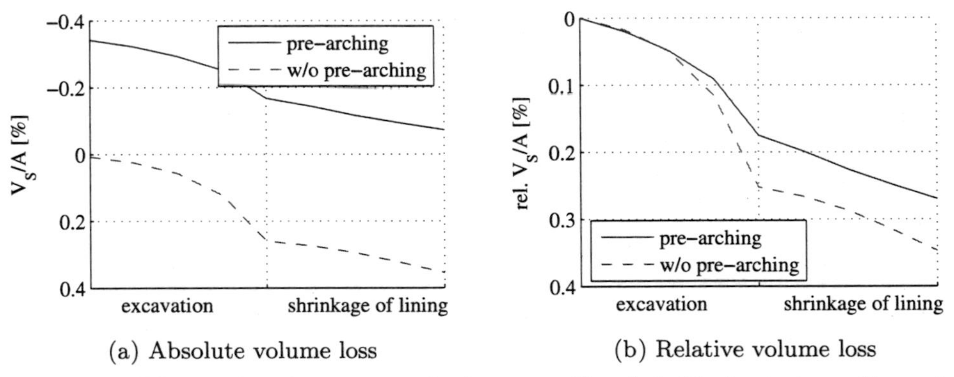

(a) Absolute volume loss (b) Relative volume loss

Fig. 6.10: Volume of settlement trough after tunnelling (rel. V_S due to tunnelling only)

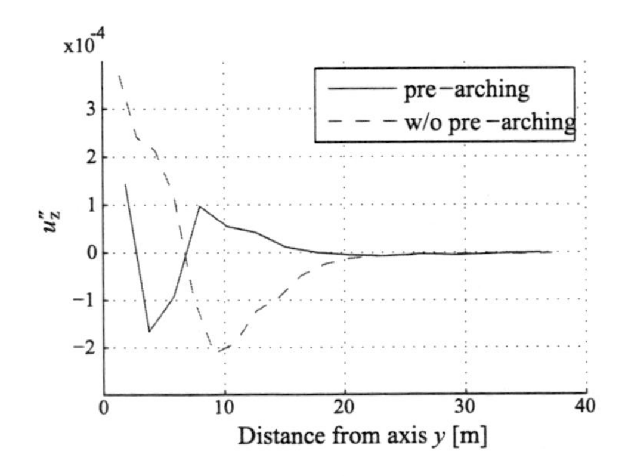

Fig. 6.11: 2nd derivation of vertical displacements u_z''.

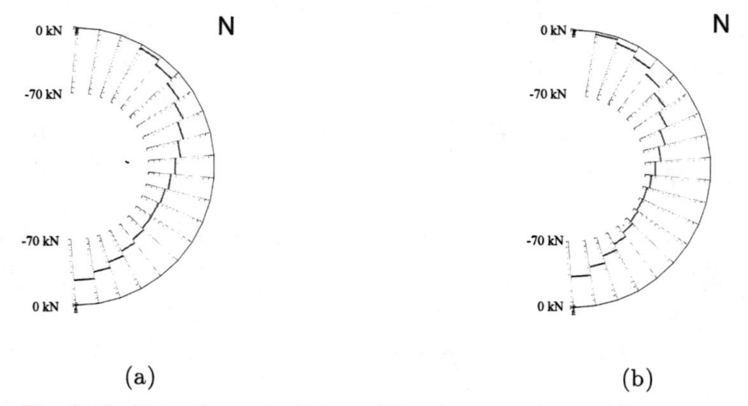

Fig. 6.12: Hoop forces in lining: (a) w/o pre-arching, (b) pre-arching.

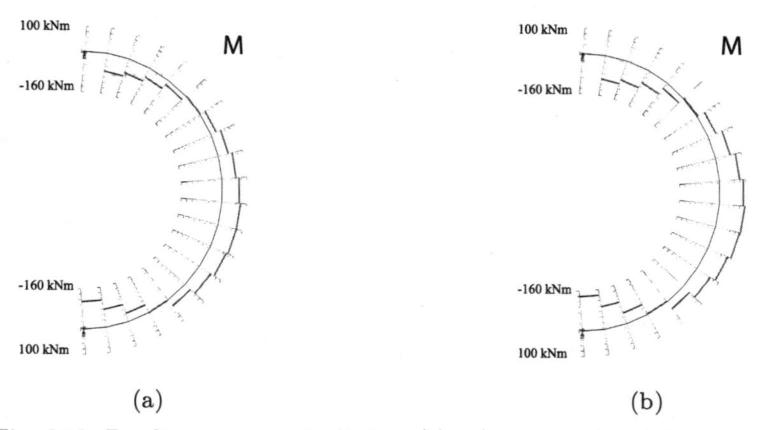

Fig. 6.13: Bending moments in lining: (a) w/o pre-arching, (b) pre-arching.

6.3 Conclusion and discussion

Pre-arching is a promising method to control ground movements. The alternative arrangement of injection zones provides the following advantages:

- ground movements are reduced up to 50 %,
- lining stresses are only slightly increased, contrary to the common used layout of grouting,
- Curvatures of surface settlement, which are a cause of building damage, are reduced.

Due to the small effects on the lining, NATM tunnelling is another area of application for this method.

The injection zones may be grouted by fracture grouting or compensation grouting. However further research is necessary to optimise grouting methods.

The right choice of the grouting zones and the grouting volume is essential to avoid settlements. The zones may be designed as follows: At first an analysis of the tunnel excavation without pre-arching is carried out. A plot of the stress trajectories may act as reference to draw the soil arch. Then the injection zones are arranged perpendicular to the arch and by means of further analyses number and grout volume of the grouting zones may be optimised.

Chapter 7

Analytical Approach

The model tests in chapter 4 showed, that the traditional concept of volume loss is not applicable for dilatant soils, as the volume of the settlement trough differs from volume loss. An extension of this widely used concept is proposed to overcome this shortcoming.

7.1 Volume of settlement trough depending on angle of dilatancy

Due to loosening of the ground the volume of the settlement trough V_S is smaller than the volume loss V_L. Thus, dilatant soil behaviour must be taken into account, if settlements are estimated using the concept of volume loss. This can be achieved by a combination of the method of PECK (section 2.4.1.3) and the method of KOLYMBAS (section 2.4.3). For this purpose the following assumptions are made:

- All displacements point toward the tunnel axis ($u_\vartheta = 0$),
- the convergence of the tunnel lining is constant along the tunnel contour,
- the settlement trough has the shape of a Gauss-curve,
- constant angle of dilatancy.

We start from the crown, where the radial displacement $u_{r,0}$ and the radius r_0 is known (Fig. 7.1). Thus, the circumferential strain $\varepsilon_{\vartheta 0} = u_{r,0}/r_0$ and the radial strain $\varepsilon_{r0} = \varepsilon_{\vartheta 0}/(\tan\psi - 1)$ are given. By integration of the radial strains the radial displacements in any distance r may be calculated:

$$u_r(r,\psi) = u_{r,0} + \int_{r_0}^{r} \frac{u_r(r,\psi)}{r(\tan\psi - 1) - 1}\,\mathrm{d}r. \tag{7.1}$$

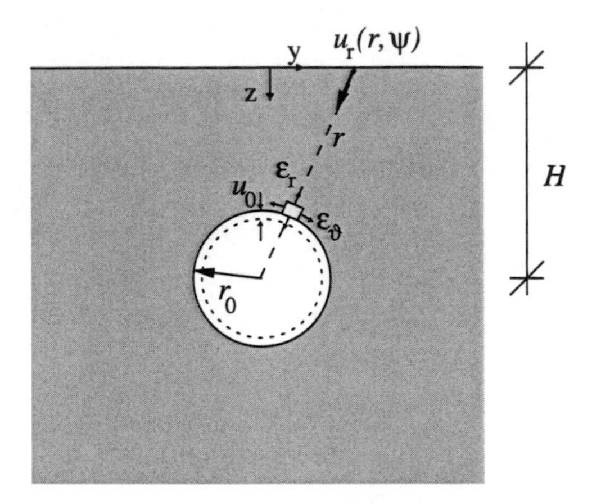

Fig. 7.1: Radial displacements

$\tan\psi$ may be replaced by b. From eq. (7.1) $u_r(r,\psi)$ is obtained as [51]:

$$u_r(r,\psi) = u_{r,0}\left(\frac{r_0}{r}\right)^{-\frac{1}{b-1}}. \tag{7.2}$$

By substituting r with $\sqrt{y^2+H^2}$ one gets the radial displacements at the surface

$$u_r(y,\psi) = u_{r,0}\left(\frac{r_0}{\sqrt{y^2+H^2}}\right)^{-\frac{1}{b-1}}. \tag{7.3}$$

The vertical displacement at the surface reads:

$$u_z(y,\psi) = u_{r,0}\left(\frac{r_0}{\sqrt{y^2+H^2}}\right)^{-\frac{1}{b-1}}\frac{H}{\sqrt{y^2+H^2}}. \tag{7.4}$$

As this solution is an extension of the method of KOLYMBAS it is termed extended KOLYMBAS solution. This equation is similar to the equation based on LAMÉ's solution (eq. 2.4.1.1). Hence, the settlement trough predicted by this equation is not realistic (cf. Fig. 7.3), but it is possible to calculate the ratio $\alpha(y,\psi)$ of the displacements with an arbitrary angle of dilatancy ψ to the displacements for the undrained case ($\psi = 0$ and $b = 0$):

$$\alpha(y,\psi) = \frac{u_z(y,\psi)}{u_z(y,\psi=0)} = \frac{u_{r,0}\left(\frac{r_0}{\sqrt{y^2+H^2}}\right)^{\frac{-1}{b-1}}\frac{H}{\sqrt{y^2+H^2}}}{u_{r,0}\left(\frac{r_0}{\sqrt{y^2+H^2}}\right)\frac{H}{\sqrt{y^2+H^2}}} = \left(\frac{r_0}{\sqrt{y^2+H^2}}\right)^{\frac{b}{1-b}} . \tag{7.5}$$

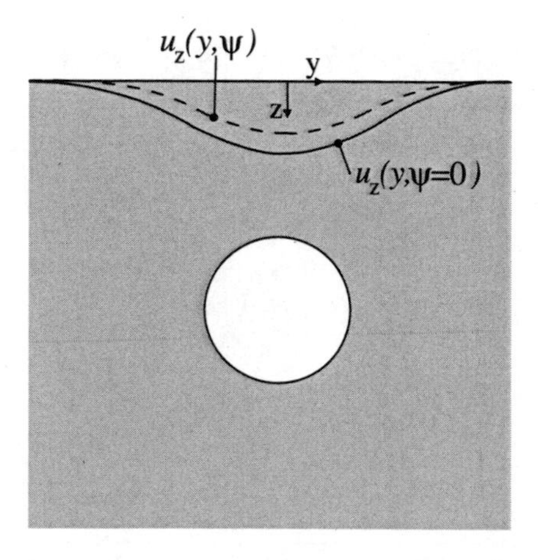

Fig. 7.2: Settlement troughs for undrained soil and soil with angle of dilatancy ψ.

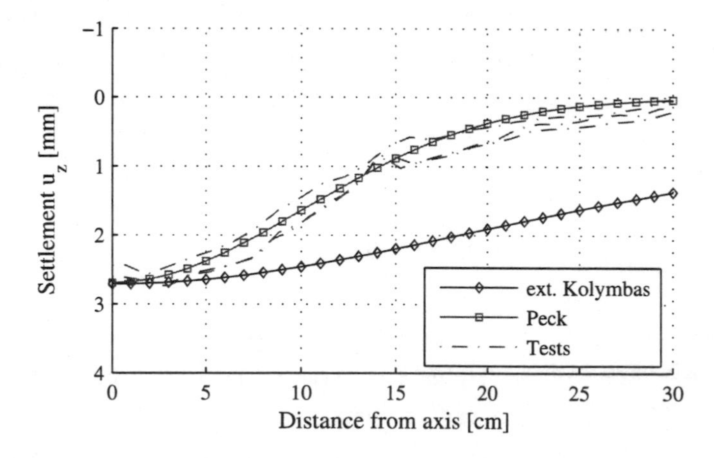

Fig. 7.3: Settlement troughs obtained with extended KOLYMBAS solution and PECKS solution compared to results of test series **OR066**. Note, that the maximum settlement has been fitted and the trough width parameter i was set to $i = 0.3H$.

As known, the Gauss-curve gives a better approximation of the settlement trough than eq. (7.4). If the Gauss-curve is multiplied by $\alpha(y,\psi)$, we obtain an improved Gauss-curve, that takes volumetric soil behaviour into account:

$$u_z(y,\psi) = u_{z,\max} \exp^{-\frac{y^2}{2i^2}} \alpha(y,\psi) = u_{z,\max} \exp^{-\frac{y^2}{2i^2}} \left(\frac{r_0}{\sqrt{y^2+H^2}} \right)^{-\frac{b}{1-b}} . \tag{7.6}$$

Integration of the settlement curve yields the volume of the settlement trough:

$$V_S(\psi) = u_{z,\max} \int_{-\infty}^{\infty} \exp^{-\frac{y^2}{2i^2}} \left(\frac{r_0}{\sqrt{y^2+H^2}} \right)^{-\frac{b}{1-b}} dy \quad . \tag{7.7}$$

The ratio of the volume of the settlement trough V_S to the volume loss V_L is termed β and reads:

$$\beta = \frac{V_S}{V_L} = \frac{u_{z,\max} \int_{-\infty}^{\infty} \exp^{-\frac{y^2}{2i^2}} \left(\frac{r_0}{\sqrt{y^2+H^2}} \right)^{-\frac{b}{1-b}} dy}{\sqrt{2\pi} i u_{z,\max}} \quad . \tag{7.8}$$

The solution of the integral in eq. (7.8) is quite long and unhandy. Therefore a design chart is presented in Fig. (7.4), where β may be obtained for any reasonable combination of ψ and $\frac{H}{D}$. The trough width parameter i was set to $i = 0.3H$ (eq. 2.9, p. 21) for the evaluation of the design chart. Parametric studies showed that the influence of the trough width parameter i on β is negligible.

7.2 Appropriateness of method

The proposed method is used to estimate the ratio β and the maximum settlements of the own model tests (chapter 4) for various initial densities $I_e = [0.4; 0.6; 0.8; 0.97]$. The H/D ratio of the apparatus reads 2.5 for all tests. In view of the fact that triaxial test under low lateral stress are very difficult and proned to errors, the dilatancy angle is obtained from numerically simulated triaxial tests (i.e. axially symmetric samples). The element tests are calculated using the hypoplastic material model with intergranular strains.

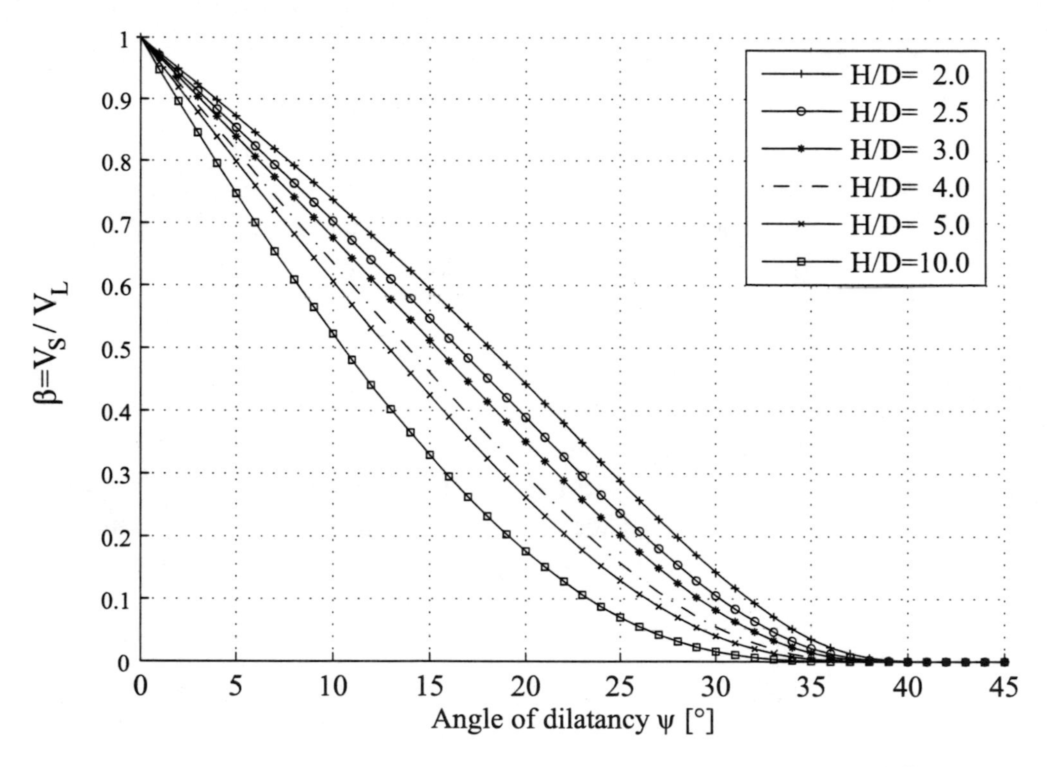

Fig. 7.4: Ratio β plotted against angle of dilatancy ψ for various ratios $\frac{H}{D}$

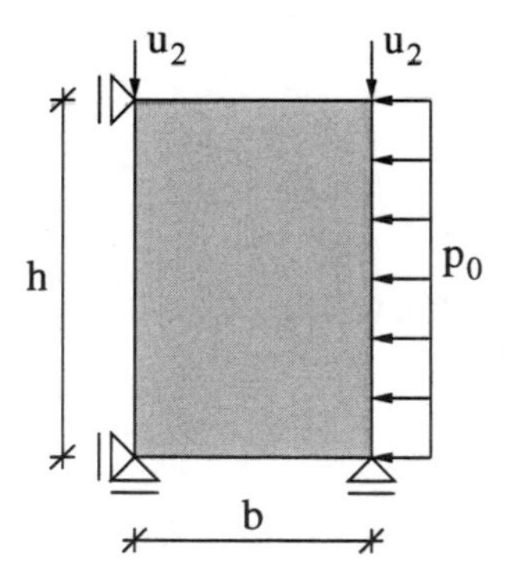

Fig. 7.5: Boundary conditions for element tests

The material parameters are the same as the ones used for the simulation of the model tests (see Tab. 5.1), which showed their appropriateness. 4-node bilinear axisymmetric elements are used, the boundary conditions are shown in Fig. (7.5). The initial stress condition is chosen according to the stresses in the model tests to $T_{11} = T_{22} = p_0 = 3$ kN/m^2. The tests are displacement-controlled, vertical displacements of $u_2 = 0.015$ m are applied to the top nodes.

As the angle of dilatancy is not constant during triaxial tests (Fig. 7.6), its determination is not straight forward. The author chose ψ as a secant modulus from the starting point to an axial strain of 4%. The latter has been chosen due to the fact that the vertical strains of the model tests are in this range. The results are presented in Tab. (7.1).

From the angle of dilatancy, the ratio β may be either calculated (eq. 7.8) or taken from the design chart in Fig. 7.4. In this example the β-values are calculated by numerical integration, the results are presented in Tab. (7.1). In Fig. (7.7) the results are compared to the test results and the results obtained with FE analyses. The analytical solution is in good agreement with both methods.

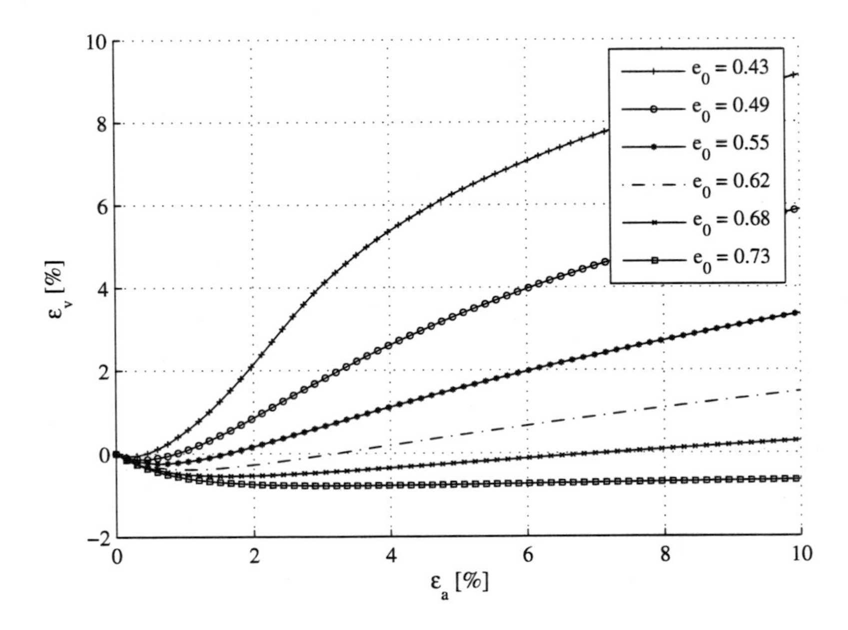

Fig. 7.6: Volumetric strain ε_v vs. axial strain ε_v, triaxial element tests with various initial void ratios.

Initial void ratio e_0 [-]	0.43	0.48	0.55	0.62
Angle of dilatancy ψ [°]	53.2	33.8	16.2	2.7
Ratio $V_S/V_L = \beta$ [-]	0.0	0.04	0.51	0.92

Tab. 7.1: Angle of dilatancy and ratio β for various initial void ratios.

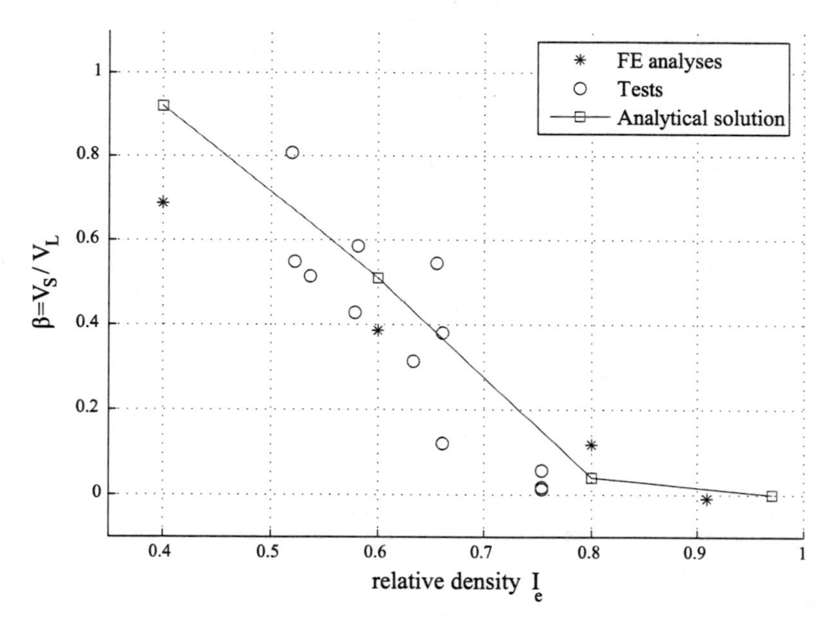

Fig. 7.7: Ratio $\beta = V_S/V_L$ obtained from analytical solution compared to settlements from test results and FE analyses after contraction of cylinder.

Now that the volume of the settlement trough may be calculated from V_L and β, it is easy to estimate the maximum settlements. The volume of the settlement trough resulting from the Gauss-curve reads:

$$V_S = \sqrt{2\pi} i u_{z,\max} \quad . \tag{7.9}$$

Assuming that the trough width parameter may be estimated with $i = 0.3H$, the maximum settlement $u_{z,\max}$ follow as

$$u_{z,\max} = \frac{V_S}{\sqrt{2\pi}0.3H} \quad . \tag{7.10}$$

The maximum settlements obtained with eq. (7.10) are compared to the results from the tests and FE analyses in Fig. 7.8.

7.3 Discussion and suggestions for application

The proposed method to estimate settlements is based upon the method of PECK. An equation for the ratio β of the volume of the settlement trough to

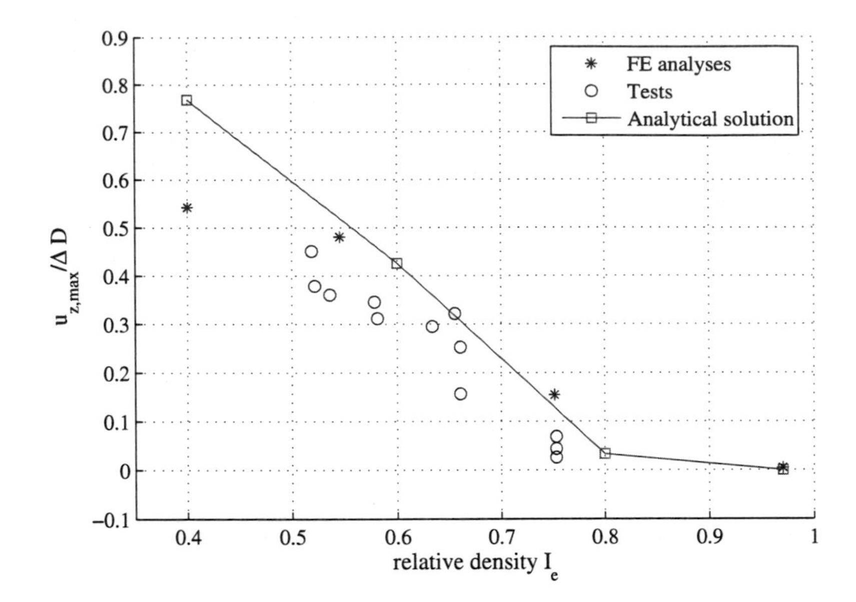

Fig. 7.8: Settlements obtained from analytical solution compared to settlements from test results and FE analyses after contraction of cylinder.

the volume loss is deduced from the method of KOLYMBAS, that takes volumetric soil behaviour into account. As the equation is based on the angle of dilatancy, both barotropy and pyknotropy are considered. Thus, by knowing the volume loss and the angle of dilatancy, the volume of the settlement trough can be estimated.

The angle of dilatancy can be assessed from ε_v vs. ε_a plots of triaxial tests. The volume loss can be either taken from empirical relationships between the volume loss V_L and stability number N (section 2.4.1.3, p. 20), from the gap method proposed by Lee and Rowe [60] (section 2.5.2.1, p. 29) or from experience. If volume loss is obtained from e.g. a nearby site, one should be aware however, that usually the volume of settlement trough V_S is reported, not the volume loss. Then, knowing the angle of dilatancy ψ of the site, the volume loss may be obtained by:

$$V_L = V_S/\beta(\psi) \quad . \tag{7.11}$$

Model tests and FE analyses confirmed the proposed method. Nevertheless further tests, especially with case studies, are recommended.

Chapter 8

Concluding remarks

The own laboratory tests and FE analyses emphasised the disparity of volume of settlement trough V_S and volume loss V_L for non-cohesive soils. Due to the volumetric effects in soil behaviour the volume of settlement trough V_S takes values from 0-100 % of volume loss V_L. This fact is neglected in many practical design methods, which therefore produce unreliable results. An analytical approach is proposed to estimate the volume of settlement trough V_S from the volume loss V_L. The approach takes into account the angle of dilatancy ψ and the ratio of overburden to tunnel diameter H/D. A comparison with laboratory tests shows the appropriateness of this method. Nevertheless, one should be aware that volume loss V_L is not a constant, but depends on geometry and other soil state variables, too.

Furthermore this thesis points out that volumetric soil behaviour of non-cohesive soils impedes the reversal of surface settlements by tail gap grouting. Shield tunnelling induces cyclic loading on the surrounding soil. Non-cohesive soil is densified by virtue of cyclic loading and, as a result, the surface is more or less unaffected from tail gap grouting in most cases.

The soil behaviour under loading-unloading cycle is a critical point in FE analyses of shield tunnelling. If the construction process is simulated properly and detailed, cyclic loading appears in the FE simulation as well. Thus, a constitutive model that is able to reproduce volumetric soil behaviour under cyclic loading is essential to obtain reliable results. A comparison of FE analyses with Mohr-Coulomb constitutive model and hypoplasticity showed the appropriateness of the latter in this respect.

The behaviour of granular soil is incrementally non-linear except for a small elastic range, not exceeding ca. 10^{-5}. In this range the stiffness is higher and approximately independent of strain. The concept of intergranular strain is an extension of hypoplasticity that takes the small strain stiffness into account and thus improves the performance of the constitutive model for small strains and after changes of direction of stress or strain path. The FE analyses of model tests have been carried out using hypoplasticity with intergranular strain. Good agreement of computed and measured results was obtained. Thus the importance of considering small strain stiffness in shield tunnelling problems is emphasised.

For this work, a new implementation of hypoplasticity with intergranular strains to the FE program ABAQUS was assembled. The implementation used a second-order Euler method as numerical integrator, which guarantees quadratic convergence of the equilibrium iteration and supplies an error control. The benefits regarding accuracy and speed were checked by means of element tests.

Shear localisation of deformation in narrow shear bands is a fundamental phenomenon and dominates the deformation patterns of soil at shield tunnelling. However, at the time being it is not possible to create FE models of shield tunnelling whose element sizes correspond to shear band width. Thus, the importance to compute the width of the localised zone correctly should be investigated and the influence of the mesh size on the results should be controlled. The rigid body failure mechanisms found by means of laboratory tests and FE analyses may help to better understand soil behaviour. Furthermore they may contribute to new analytical approaches to predict the surface settlements.

Finally, a new arrangement of grouting zones for compensation grouting is proposed, which is termed pre-arching. By arranging the grouting zones perpendicular to a virtual arch, arching may be mobilised previous to tunnelling. Thus, less soil movement is necessary to obtain equilibrium after tunnelling and hence less surface settlements occur. The method is tested by means of FE analyses, which give promising results. Further research is necessary to investigate feasibility for practical application. 3D FE analyses should be carried out to investigate the arch effect more detailed. Grout behaviour in case of inclined grouting zones should be studied. In case of successful application the method may be extended to NATM tunnelling, as the additional loading of the lining is by far smaller than due to horizontal grouting layers above the crown.

References

[1] A.A.M. Abu-Krisha. *Numerical modelling of TBM tunneling in consolidated clay.* PhD thesis, University of Innsbruck, 1998.

[2] T. Addenbrooke and D.M. Potts. Twin tunnel construction - ground movements and lining behaviour. In R.J. Mair and R.N. Taylor, editors, *International Symposium on Geotechnical Aspects of Underground Construction in Soft Ground*, pages 441–446, Rotterdam, 1996. Balkema.

[3] T.I. Addenbrooke. Recent advances in modelling of ground movements due to tunneling. *Ground Engineering*, 28(7):40–44, September 1995.

[4] Herrenknecht AG. Through soft ground with earth pressure. Brochure, 2001.

[5] Herrenknecht AG. http://www.herrenknecht.de, February 2005.

[6] Hirokazu Akagi. Computational Simulation of Shield Tunneling in Soft Ground. *Memoirs of the School of Science & Engineering*, (58), 1994.

[7] W. Angerer. Numerical Simulation in Shield Tunneling. Master's thesis, Institute of Geotechnical and Tunnelling Engineering, University of Innsbruck, 2003.

[8] J. H. Atkinson and D. M. Potts. Subsidence above shallow tunnels in soft ground. *Journal of the Soil Mechanics and Foundations Division*, 103(4):307–325, April 1977.

[9] P.B. Attewell. *Large ground movements and structures*, chapter Ground movements caused by tunneling in soil, pages 812–948. Pentech Press, London, 1978.

[10] P.B. Attewell and J.P. Woodmann. Ground movement caused by tunnelling in soil. *Ground Engineering*, 15(8):32–41, 1982.

[11] C.E. Augarde. *Numerical Modelling of Tunnelling Processes for Assessment of Damage to Buildings.* PhD thesis, Keble College, 1997.

[12] K.J. Bakker and W. van Schelt. Predictions and a monitoring scheme with respect to the boring of the second heinenoord tunnel. In R.J Mair and R.N. Taylor, editors, *Geotechnical Aspects of Underground Construction in Soft Ground*, pages 459–464, Rotterdam, 1996. Balkema.

[13] K.-J. Bathe. *Finite-Elemente-Methoden.* Springer, 2002.

[14] E. Bauer. Calibration of a comprehensive hypoplastic model for granular materials. *Soils and Foundations*, 36(1):169–183, 1996.

[15] J. Bochert. Modellversuche zur Ringspaltverpressung. Technical report, University of Dresden, 2005.

[16] M. Bolton. *A Guide to Soil Mechanics.* The Mc Millian Press Ltd., 1979.

[17] W. Broere and R.B.J. Brinkgreve. Phased simulation of a tunnel boring process in soft soil. In Mestad, editor, *Numerical Methods in Geotechnical Engineering*, pages 529–536, 2002.

[18] B.B. Broms and H. Bennermark. Stability of clay at vertical openings. In *Proc. ASCE: 93 SMI*, pages 71–94, 1967.

[19] J. Burland, J.R. Standing, and F.M. Jardine. Assessing the risk of building damage due to tunnelling - lessons from the jubilee line extension, London. In *Int. Conference on Soil Structure Interaction in Urban Civil Engineering*, pages 11 – 38. ETH Zürich, March 2002.

[20] T.B. Celestino and A.P.T. Ruiz. Shape of settlement troughs due to tunnelling through different types of soft ground. *Felsbau*, 16(2): 118–121, 1998.

[21] . Modelling in geotechnics. Technical report, ETH Zürich, Institute of Geotechnical Engineering, http://geotec4.ethz.ch/mig/, 2003.

[22] G.W. Clough and B. Schmidt. *Soft Clay Engineering*, chapter Design and Performance of Excavations and Tunnels in Soft Clay, pages 567–634. Elsevier, 1981.

[23] E.J Cording, W.H. Hansmire, H.H. MacPherson, P.A. Lenzini, and A.D. Vonderohe. Displacements around tunnels in soil. Technical report, University of Illinois, Dept. of Transportation, 1976.

[24] R. Craigg. Discussions at the meeting of the british tunnelling society. *Tunnels and Tunnelling*, 7:61–65, 1975.

[25] E.H. Davis, M.J. Gunn, R.J. Mair, and H.N. Seneviratnes. The stability of shallow tunnels and underground openings in cohesive material. *Géotechnique*, 30(4):397–416, 1980.

[26] Z. Eisenstein. Geotechnical challanges in soft ground tunnelling – examples from significant projects. In Vanicek et al., editor, *Proc. XIII ECSMGE*, volume 3, pages 779–782, Prague, 2003.

[27] E. Falk. *Bodenverbesserung durch Feststoffeinpressung mittels hydraulischer Energie.* PhD thesis, Technical University of Vienna, 1998.

[28] W. Fellin and A Ostermann. Consistent tangent operators for constitutive rate equations. *International Journal for Numerical and Analytical Methods in Geomechanics*, 26:1213–1233, 2002.

[29] W. Fellin and A Ostermann. Umat. http://www2.uibk.ac.at/geotechnik/res/fehypo.html, 2002.

[30] H. Fromm. Experimentelle Überprüfung von Oberflächensetzungen infolge Ringspaltes und seiner Verpressung. Master's thesis, Institute of Geotechnical and Tunnelling Engineering, University of Innsbruck, 2002.

[31] C. Gonzalez and C. Sagaseta. Patterns of soil deformations around tunnels. Application to the extension of madrid metro. *Computers and Geotechnics*, 28:445–468, 2001.

[32] R.J. Grant and R.N. Taylor. Tunnelling-induced ground movements in clay. In *Proc. Instn. Civ. Engrs. Geotech. Engineering*, volume 143, pages 43–55, 2000.

[33] G. Gudehus. Attractors, percolation thresholds and phase limits of granular soil. In Behringer and Jenkins, editors, *Powders and Grains*, pages 169–183. Balkeema, 1997.

[34] M.J. Gunn. The prediction of surface settlement profiles due to tunnelling. In Thomas Telford, editor, *Predictive Soil Mechanics*, Oxford, 1993. Wroth Memorial Symposium.

[35] E. Hairer, S.P. Nørsett, and G. Wanner. *Solving Ordinary Differential Equations I. Nonstiff Problems.* Springer, Berlin, 1993.

[36] W.H. Hansmire and E.J. Cording. Soil tunnel test section: case history summary. *Journal of Geotechnical end Geoenvironmental Engineering*, 111(11):1301–1320, 1984.

[37] T. Hashimoto, Nagaya J., and Takahiro K. Prediction of ground deformation due to shield excavation in clayey soils. *Soils and Foundation*, 39(3):53–61, June 1999.

[38] I. Herle. Hypoplastizität und Granulometrie einfacher Korngerüste. Heft 142, Institut für Bodenmechanik und Felsmechanik, University of Karlsruhe, 1997.

[39] I. Herle. Numerical implementation of the intergranular strain model. Internal paper, October 2000.

[40] I. Herle and G. Gudehus. Determination of parameters of a hypoplastic constitutive model from properties of grain assemblies. *Mechanics of Cohesive frictional materials*, 4(5):461–486, 1999.

[41] H.M. Hügel. Prognose von Bodenverformungen. Heft 136, Institut für Bodenmechanik und Felsmechanik, University Fridericiana, Karsruhe, 1995.

[42] Hibbitt, Karlsson & Sorensen, Inc. *Abaqus/Standard User's Manual VOL I-III*, 2002.

[43] T. Hirata, Y. Aritome, T. Mishima, H. Eguchi, and T. Hashimoto. Behaviour of soft clayey ground in EPB shield drivings. In *Proceedings of 18th JSSMFE*, pages 1339–1342, 1983.

[44] Babendererde Ingenieure. Projektmappe, 2003.

[45] S. Jancsecz, W. Frietzsche, J. Breuer, and K.R. Ulrichs. *Tunnelbau 2001*, volume 25, chapter III. Minimierung von Senkungen beim Schildvortrieb am Beispiel der U-Bahn Düsseldorf, pages 165–214. Glückauf, Essen, 2001.

[46] R.J. Jardine, M.J. Symes, and J.B. Burland. The measurement of soil stiffness in the triaxial apparatus. *Géotechnique*, 34:323–340, 1984.

[47] T. Kasper and G. Meschke. A 3D finite element simulation model for TBM tunnelling in soft ground. *Int. J. Numer. Anal. Meth. Geomech.*, 28:1441–1460, 2004.

[48] D. König. *Beanspruchung von Tunnel- und Schachtausbauten in kohäsionslosem Lockergestein unter Berücksichtigung der Verformungen im Boden.* PhD thesis, University of Bochum, 1994.

[49] D. Kolymbas. Ein nichtlineares visko-plastisches Stoffgesetz für Böden. Heft 77, Institut für Bodenmechanik und Felsmechanik, University Fridericiana, Karsruhe, 1978.

[50] D. Kolymbas. *Eine konstitutive Theorie für Böden und andere körnige Stoffe.* Habilitation, University of Karlsruhe, 1988.

[51] D. Kolymbas. *Geotechnik – Tunnelbau und Tunnelmechanik.* Springer, August 1998.

[52] D. Kolymbas, M. Mähr, and I. Herle. Tail gap grouting. In *Proceedings of ACUUS 2002 - Urban Underground space: a Resource for Cities*, Turin, 2002.

[53] K. Komiya. Finite element modelling of TBM excavation and advancement processes in soft ground. In E. Onate and D.R.J. Owen, editors, *VII International Conference on Computational Plasticity*, 2003.

[54] K. Komiya, K. Soga, H. Akagi, T. Hagiwara, and Bolton M. Finite element modelling of excavation and advancement processes of a shield tunnelling machine. *Soils and Foundation*, 1999.

[55] N. Kovacevic, H.E. Edmonds, R.J. Mair, and K.G. Higgins. Numerical modelling of the NATM and compensation grouting trials at Redcross way. In Mair & Taylor, editor, *Geotechnical Aspects of Underground Construction in Soft Ground*, pages 553–559. Balkeema, 1996.

[56] E.A. Kwast and J.W. Plekkenpol. Ground deformations of stresses in the surrounding and ground pressures on the tunnel lining, k100-w-009. Technical report, 1996.

[57] J. LANIER, C. DI PRISCO, AND R. NOVA. Induced anisotropy in hostun sand: experiments and theoretical analysis. *Rev. Franc. Geotech.*, (57): 59–74, October 1991.

[58] A. LAUDAHN. *An Approach to 1g Modelling in Geotechnical Engineering with Soiltron.* PhD thesis, Institute of Geotechnical and Tunnelling Engineering, University of Innsbruck, 2004.

[59] K.M. LEE. *Prediction of ground deformation resulting from shield tunnelling in soft clays.* PhD thesis, The University of Western Ontario, 1989.

[60] K.M. LEE AND R.K. ROWE. An analysis of three-dimensional ground movements: the thunder bay tunnel. *Canadian Geotechnic Journal*, 28: 25–41, 1991.

[61] K.M. LEE, R.K. ROWE, AND K.Y. LO. Subsidence owing to tunnelling. I. estimating the gap parameter. *Canadian Geotechnic Journal*, 29:929–940, 1992.

[62] S.R. MACKLIN. The prediction of volume loss due to tunneling in over consolidated clay based on heading geometry and stability number. *Ground Engineering*, 32(4):30–34, April 1999.

[63] R. J. MAIR, R. N. TAYLOR, AND A. BRACEDGIRDLE. Subsurface settlement profiles above tunnels in clays. *Géotechnique*, 43(2):315–320, 1993.

[64] R.J. MAIR, M.J. GUNN, AND M.P O'REILLY. Ground Movements around Shallow Tunnels in Soft Clay. In *Proceedings of the Tenth Int. Conf. on Soil Mechanics and Foundation Engineering*, volume 1, pages 323–328. Balkema, 1981.

[65] R.J. MAIR AND R.N. TAYLOR. Bored tunneling in the urban environment. In *14th Int. Conf. SMFE*, Hamburg, 1997.

[66] M.A.M. MANSOUR. *Three-Dimensional Numerical Modelling of Hydroshield Tunnelling.* PhD thesis, University of Innsbruck, 1996.

[67] P. MELIX. Modellversuche und Berechnungen zur Standsicherheit oberflächennaher Tunnel. Heft 103, Institut für Bodenmechanik und Felsmechanik, University Fridericiana, Karsruhe, 1987.

[68] M. MÄHR AND I. HERLE. Volume loss and soil dilatancy. *Rivista Geotechnica Italiana*, (4):32–40, 2004.

[69] T. NAKAI, L.M. XU, AND YAMAZAKI H. Prediction of surface settlement profiles due to tunnel excavation: Model tests and numerical analyses. In Pande & Pietruszczak, editor, *Numerical Models in Geomechanics - NUMOG V*, pages 495–500, Rotterdam, 1995. Balkema.

[70] K. NÜBEL. Experimental and Numerical Investigation of Shear Localization in Granular Material. Heft 156, Institut für Bodenmechanik und Felsmechanik, University Fridericiana, Karsruhe, 2002.

[71] K. NÜBEL AND A. NIEMUNIS. umatibf.f, 1999. Fortran Implementierung des hypoplastischen Stoffgesetzes mit intergranularer Dehnung in Abaqus.

[72] K. NÜBEL AND V. WEITBRECHT. Visualization of Localisation in grain skeletons with Particle Image Velocimetry. *Journal of testing and evaluation*, 30(4):322–329, 2002.

[73] A. NIEMUNIS AND I. HERLE. Hypoplastic model for cohesionless soils with elastic strain range. *Mechanics of Cohesive-Frictional Materials*, 2:279–299, 1997.

[74] A. NIEMUNIS, K. NÜBEL, AND CH. KARCHER. The consistency conditions for density limits of hypoplastic constitutive law. In J. Tejchman, editor, *Publications of TASK*, volume 4, pages 412–420, 2000.

[75] T. NOMOTO, S. IMAMURA, T. HAGIWARA, O. KUSAKABE, AND N. FUJII. Shield tunnel construction in centrifuge. *Journal of Geotechnical end Geoenvironmental Engineering*, 125(4):289–300, April 1999.

[76] T. NOMOTO, K. MITO, AND S. IMAMURA. Centrifuge modelling of construction processes of shield tunnel. In Mair & Taylor, editor, *Geotechnical Aspects of Underground Construction in Soft Ground*, pages 567–572, Rotterdam, 1996. Balkema.

[77] J. OHDE. Zur Theorie der Druckverteilung im Baugrund. *Bauingenieur*, 20:451–459, 1938.

[78] M.P. O'REILLY AND B.M. NEW. Settlements above tunnels in the United Kingdom — their magnitude and prediction. In *Proc. Tunnelling Symp '82*, London, 1982. Institution of Mining and Metallurgy.

[79] M. Panet and A. Guenot. *Tunnelling 82*, pages 197–204. The Institution of Mining and Metallurgy, London, 1982.

[80] R. B. Peck. Deep excavations and tunneling in soft ground. State-of-the-Art report. In *Proceedings of the 7th International Conference on Soil Mechanics and Foundation Engineering*, volume State-of-the-Art Volume, pages 225–290, Mexico City, 1969.

[81] D.M. Potts and L. Zdravkovic. *Finite Element analysis in geotechnical engineering: application.* Thomas Telford, London, 2001.

[82] M. Raffel, C. Willert, and Kopenhans J. *Particle Image Velocimetry, A practical guide.* Springer-Verlag Berlin, 1998.

[83] W.J. Rankin. Ground movements resulting from urban tunnelling: predictions and effects. *Engineering Geology od Underground Movements*, (Special Publication No. 5):79–92, 1988.

[84] W. Rinawi. Particle Image Velocimetry (PIV) applied on triaxial tests. Master's thesis, Institute of Geotechnical and Tunnelling Engineering, University of Innsbruck, 2004.

[85] D. Roddeman. FEM-implementation of hypoplasticity. Technical report, Institute of Geotechnical and Tunnelling Engineering, University of Innsbruck, 1997.

[86] K.H. Roscoe. The influence of strains on soil mechanics. *Géotechnique*, 20(2):129–170, 1970.

[87] R.K. Rowe and G.J. Kack. A theorethical examination of the settlements induced by tunnelling: four case histories. *Canadian Geotechnical Journal*, 20:299–314, 1983.

[88] R.K. Rowe and K.Y. Lo. A method of estimating surface settlement above tunnels constructed in soft ground. *Canadian Geotechnical Journal*, 20:11–20, 1983.

[89] C. Sagaseta. Analysis of undrained soil deformation due to groundloss. *Géotechnique*, 37(3):301–320, 1987.

[90] C. Sagaseta. Discussion: Analysis of undrained soil deformation due to ground loss. *Géotechnique*, 37(3):301–320, 1987.

[91] B. SCHMIDT. *Settlements and ground movements associated with tunnelling in soil.* PhD thesis, University of Illinois,Urbana, 1969.

[92] J.S SHARMA AND M.D. BOLTON. A new technique for simulation of collapse of a tunnel in a drum centrifuge. Technical report, University of Cambridge, 1996.

[93] Z. SIKORA. Hypoplastic flow of granular materials. Heft 123, Institut für Bodenmechanik und Felsmechanik, University Fridericiana, Karsruhe, 1992.

[94] U. STOFFERS. *Berechnungen und Zentrifugen-Modellversuche zur Verformungssabhängigkeit der Ausbaubeanspruchung von Tunnelausbauten in Lockergestein.* PhD thesis, University of Bochum, 1987.

[95] C. TAMAGNINI. Two-dimensional FE modelling of ground deformation induced by shield tunnelling: the role of gap geometry. Lecture at ALERT-Geomaterials Workshop, October 2003.

[96] J. TEJCHMAN. *Modelling of shear localisation and autogeneous dynamic effects in granular bodies.* PhD thesis, Institut für Bodenmechanik und Felsmechanik, University of Karlsruhe, 1996.

[97] K. THERZAGHI. *Theoretical Soil Mechanics.* John Wiley & Sons, 1943.

[98] C. TRUESDELL. Hypoelasticity. *Journal of rational mechanics and analysis*, 4:83–133, 1955.

[99] C. TRUESDELL AND W. NOLL. *The Non-Linear Field Theories of Mechanics.* Springer, 1992.

[100] I. VARDOULAKIOS AND B. GRAF. Calibration on constitutive models for granular materials using data from biaxial experiments. *Géotechnique*, 5:57–78, 1981.

[101] A. VERRUIJT AND J.R. BOOKER. Surface settlements due to deformation of a tunnel in an elastic half plane. *Géotechnique*, 46(4):753–756, 1996.

[102] P.-A. VON WOLFFERSDORFF. A hypoplastic relation for granular materials with a predifined limit state surface. *Mechanics of Cohesive-Frictional Materials*, pages 251–271, 1996.

[103] D.J. White, W.A. Take, and M.D. Bolton. A deformation measurement system for geotechnical testing based on digital imaging, close-range photogrammetry, and PIV image analysis. pages 539–542. 15th International Conference on Soil Mechanics and Geotechnical Engineering, Istanbul., Balkeema, 2001.

[104] D.J. White, W.A. Take, and M.D. Bolton. Measuring soil deformation in geotechnical models using digital images and PIV analysis. In *Proceedings of the 10th International conference on Computer Methods and Advances in Geomechanics. Tucson, Arizona.*, pages 997–1002, Rotterdam, 2001. Balkema.

[105] O.C. Zienkiewitsch and R.L. Taylor. *The Finite Element Method.* MacGraw-Hill Book Company, 1994.

List of symbols and mathematical operations

Symbol	Meaning
Scalars	
c	spring stiffness
c_u	undrained cohesion
d_{50}	average grain size
e	void ratio
f_b	factor of barotropy
f_d	factor of pyknotropy
f_e	factor of pyknotropy
h	overburden surface - tunnel crown
h_s	granular hardness
i	distance from tunnel centreline to the point of inflection of settlement through
n	proportional constant
m_R, m_T	intergranular strain material parameters
m	number of elements
p	pore pressure
p_a	pressure outside a pipe
p_i	pressure inside a pipe
q_u	unconfined compressive strength

Symbol	Meaning
Scalars	
r_a	outer radius of a pipe
r_e	stress dependent relative density
r_i	inner radius of a pipe
r	radius
r_0	initial tunnel radius
u_x, u_y, u_z	displacements in x-, y-, z-direction
$u_{z,max}$	maximum settlement
Δq_{tol}	convergence criteria for nodal displacement
Δr_{tol}	convergence criteria for residual force
A	tunnel cross section; area
D	tunnel diameter
E	Young's modulus
F	factor of hypoplasticity
G	shear strength
K	Earth pressure coefficient; parameter to calculate i
H	overburden surface - tunnel axis
I_L	liquidity index
L	shield length
LF	load factor
N	stability number, length of interrogation cell
N_{TC}	stability number at collapse
M	width of interrogation cell
R	max. norm of intergranular strains
V_L	volume loss
V_S	volume of settlement trough
V_P	volume loss due to pitching

Symbol	Meaning
Scalars	
U	coefficient of uniformity
U_{3D}^*	over-excavation due to 3D movements ahead of the tunnel face
$ATOL$	threshold value
ELA	threshold of elastic nucleus
EST	estimated error
TOL	user supplied tolerance
α	stress release ratio; material parameter
$\bar{\alpha}$	stress release ratio
β	hypoplastic material parameter
β_R	intergranular strain material parameter
γ	density
δ	ovality, clearance required for erection of lining
δ_{2D}	deformation in 2D calculation
δ_{3D}	deformation in 3D calculation
ε_{r0}	strain in radial direction
$\varepsilon_{\vartheta 0}$	strain in tangential direction
λ	pitching angle
μ	Poisson's ratio
ρ	relative ovalization, normalised magnitude of intergranular strains
σ_∞	initial stress state
σ_r	stress in radial direction (cylindrical coordinates)
σ_t	supporting pressure at excavation face
σ_v	stress in vertical direction
σ_ϑ	stress in tangential direction (cylindrical coordinates)

Symbol	Meaning
Scalars	
σ_z	stress in z-direction (cylindrical coordinates)
χ	material parameter of intergranular strains
ω	gap due to workmanship
Δ	thickness of tailskin
1st order tensor	
$\mathbf{f}$	volume forces
$\mathbf{g}$	gravitational force
$\mathbf{n}$	outward normal vector on plane
$\mathbf{q}$	nodal displacements
$\mathbf{r}$	residual force
$\mathbf{t}$	stress vector on plane
$\mathbf{u}$	displacement
2nd order tensor	
$\mathbf{B}$	matrix operator
$\mathbf{D}$	symmetric part of the velocity gradient
$\mathbf{K}$	stiffness matrix
$\mathbf{L}$	operator
$\mathbf{N}$	shape functions
$\mathbf{Q}$	additional state variables
$\mathbf{R}$	cross correlation array
$\mathbf{S}$	intergranular strains

Symbol	Meaning
2nd order tensor	
$\mathbf{T}$	Cauchy stresses
$\mathbf{V}$	standard basis vector
$\mathbf{W}$	spin tensor
$\mathbf{Z}$	incidence matrix
$\boldsymbol{\varepsilon}$	strain
$\boldsymbol{\sigma}$	Cauchy stress
4th order tensor	
$\mathcal{I}$	fourth order unit tensor
$\mathcal{L}$	constitutive tensor (hypoplasticity)
$\mathcal{M}$	Jacobian $\frac{\partial \sigma_{ij}}{\partial \varepsilon_{kl}}$

Operation
$\mathbf{AB} = A_{ik}B_{kj}$
$\mathbf{A} : \mathbf{B} = A_{ij}B_{ij}$
$\mathcal{A}\mathbf{B} = A_{ijkl}B_{kl}$
$\mathrm{tr}\,\mathbf{A} = A_{ii}$
$\lVert\mathbf{A}\rVert = \sqrt{A_{kl}A_{kl}}$
$\mathrm{tr}\,(\mathbf{AB}) = A_{ij}B_{ij}$

Appendix A

Verification of umatvig.f

As FE-calculations with high computational costs have been carried out in this work, an effective and reliable implementation of the constitutive equations was essential. Therefore UMATVIG.F was programmed to implement hypoplasticity with intergranular strain (see 3.2.2) to ABAQUS. UMATVIG.F uses the consistent tangent operator, which guarantees quadratic convergence of Newton's method. Furthermore an error control, which controls the local errors in sub stepping, is implemented according to [28].

In order to verify the implementation (UMAT) of hypoplasticity with intergranular strains oedometric and triaxial element tests have been performed. The implementation of hypoplasticity with intergranular strains using the Euler-forward integration scheme written by Nübel and Niemunis [71], which are referred to as UMATIBF1.F, has been used as benchmark.

To show the sensitivity of the results regarding the increment size of ABAQUS, four calculations with various increment sizes (0.1—0.01—0.001—0.0001) have been carried out for each test and each UMAT. The increment size has been kept constant during the calculations. The sub step size is set to $1 \cdot 10^{-4}$ (=R), at the beginning of each increment and to $1 \cdot 10^{-5}$ (=$0.1R$) at strain reversals. In UMATVIG.F the local error tolerances of the stresses is set to 10^{-3} kN/m^2 and the one of intergranular strains to 10^{-5}.

The parameters in Tab. A.1 according to [39] have been used for all element tests of this chapter. They correspond to Hochstetten sand.

ϕ_c [°]	h_s [MPa]	n	e_{d0}	e_{c0}	e_{i0}	α	β
33	1000	0.25	0.55	0.95	1.05	0.25	1.50

R	m_R	m_T	β_r	χ
$1 \cdot 10^{-4}$	5.0	2.0	0.50	6.0

Tab. A.1: Hypoplastic and intergranular strain parameters of Hochstetten sand

A.1 Oedometric tests

The oedometric element tests are strain controlled, the void ratio at the beginning is chosen as 0.7. The initial stress conditions are isotropic: $T_{11} = T_{22} = T_{33} = -0.1$ MPa. All initial values S_{ij} of the intergranular strain tensor except $S_{22} = 0.0001$ are set to zero. The test starts with a compression of $\Delta\varepsilon_{22}^{(1)} = 0.01$, then unloading of $\Delta\varepsilon_{22}^{(2)} = -0.002$ follows and finally a loading of $\Delta\varepsilon_{22}^{(3)} = 0.007$ is applied.

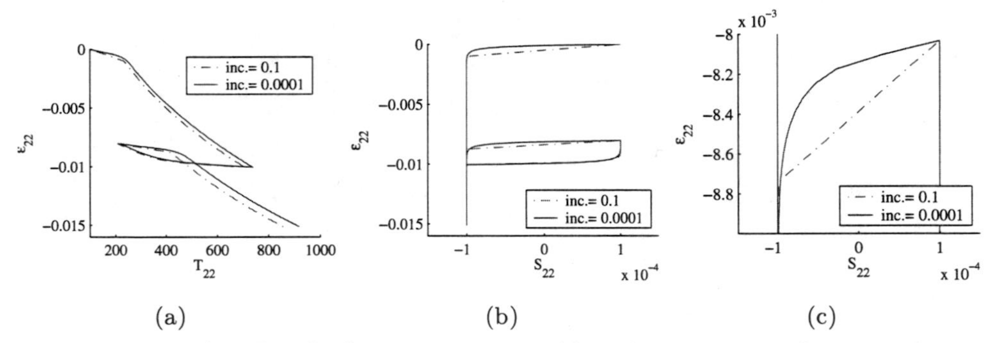

(a) (b) (c)

Fig. A.1: Results from UMATIBF1.F with various increment sizes.

Considering the results obtained using UMATIBF1.F in Fig. (A.1), the sensitivity of this implementation on a variation of the increment sizes is obvious. The stress at the end of the test differs up to 7 % ($\sigma_{22}^{(0.01)} = 852.63$ kN/m^2, $\sigma_{22}^{(0.0001)} = 917.37$ kN/m^2). The inexact calculation of the intergranular strain (see Fig. A.1b and c) causes this error, as the stiffness depends on $\hat{\mathbf{S}}\hat{\mathbf{S}}$. At a strain of -0.0099 one gets $S_{22}^{(0.01)} = 3.5371e-05$ using an increment size of 0.01, while using an increment size of 0.0001 on gets $S_{22}^{(0.0001)} = 3.7984e-05$.

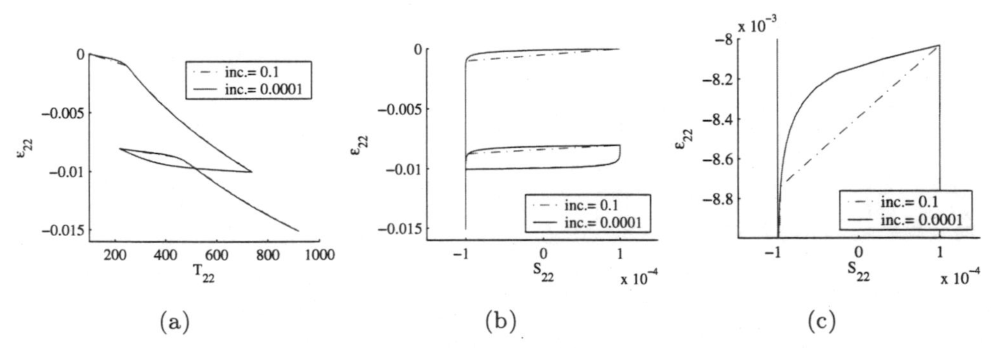

(a) (b) (c)

Fig. A.2: Results from UMATVIG.F with various increment sizes.

The results of calculations using UMATVIG.F with large increments are evidently more reliable (see Fig A.2 and Fig. A.3). The axial stresses at the end of the oedometric test differ only slightly: $\sigma_{22}^{(0.01)} = 923.2\ \text{kN/m}^2, \sigma_{22}^{(0.0001)} = 919.4\ \text{kN/m}^2$, i.e. 0.4%.

While the accuracy of the calculation improves remarkably using UMATVIG.F, the change of computational costs is admissible (Tab. A.2). Keeping in mind that UMATVIG.F provides higher accuracy for small increments than UMATIBF1.F, the computational costs for the same accuracy are smaller (Fig. A.3).

	CPU-time [s]	
	UMATIBF1.F	UMATVIG.F
inc.=0.1	13.800	18.25
inc.=0.01	35.89	57.32
inc.=0.001	113.20	295.41
inc.=0.0001	3586.80	5543.80

Tab. A.2: Comparison of computational costs (CPU-time)

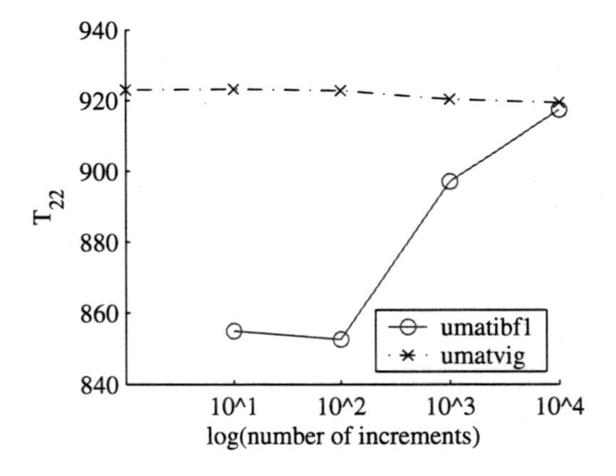

Fig. A.3: Stress $\mathbf{T}_{22}$ at the end of oedometric test for various increment numbers comparing UMATIBF1.F and UMATVIG.F

A.2 Triaxial tests

The triaxial element tests have been carried out strain-controlled, starting with an initial isotropic stress condition $T_{11} = T_{22} = T_{33} = -0.1$ MPa and a void ratio of $e_0 = 0.7$. All initial values S_{ij} of the intergranular strain tensor except $S_{22} = 0.0001$ are set to zero. The test starts applying a compression of $\Delta\varepsilon_{22}^{(1)} = 0.01$, then unloading of $\Delta\varepsilon_{22}^{(2)} = -0.001$ is applied and finally a loading of $\Delta\varepsilon_{22}^{(3)} = 0.081$ ends the simulation.

While UMATIBF1.F seems to be insensitive to a change of the increment size regarding the accuracy of the stresses, it reacts very sensitive regarding the volumetric strain (see A.4). The error of the volumetric strain goes up to 8% as can be seen in Fig. (A.6). This sensitivity results from the inaccurate calculation of the intergranular strain. Although the curves Fig. (A.4a) show perfect coincidence, the deviations are big enough to provoke inaccurate strain results.

UMATVIG.F gives better results due to an error control of the intergranular strains (see Fig. A.5). It allows accurate calculation of stresses and strains within reasonable computational costs, as can be seen in Fig. (A.6).

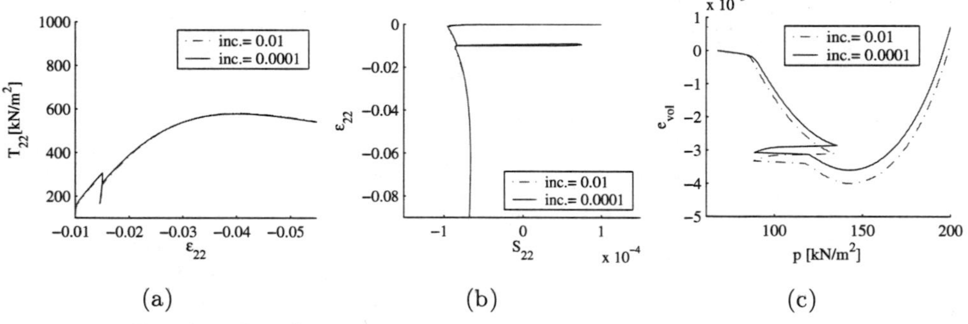

(a) (b) (c)

Fig. A.4: Results using UMATIBF1.F with various increment sizes.

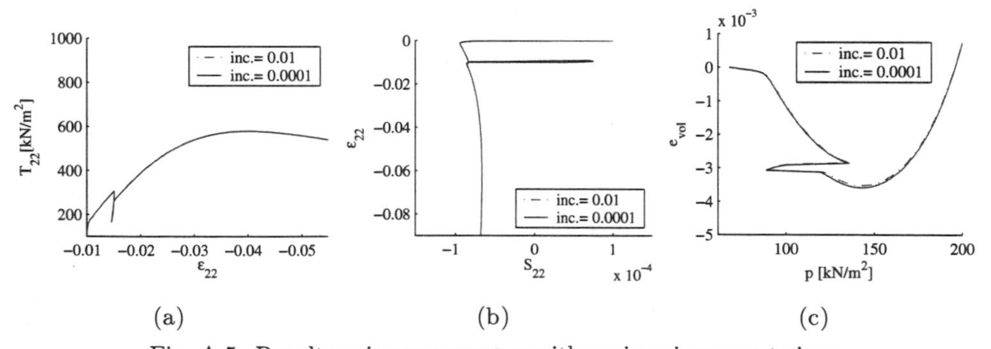

(a) (b) (c)

Fig. A.5: Results using UMATVIG.F with various increment sizes.

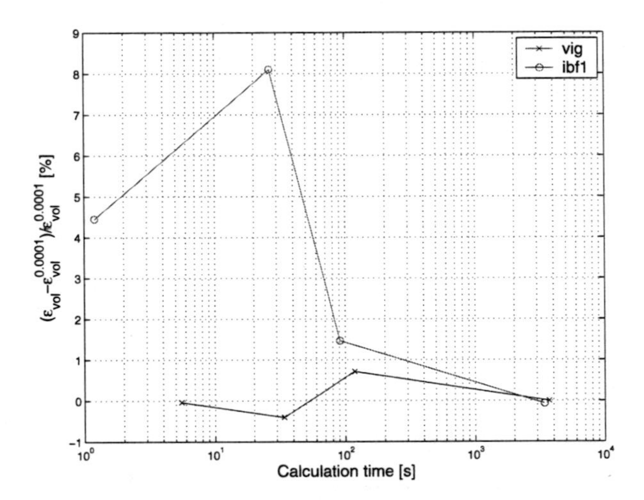

Fig. A.6: Error of ε_{vol} against calculation time of element test.

Appendix B

Numerical parameters of calculations

In non-linear calculations a huge set of numerical parameters governs the calculation. The right choice of the parameters is of paramount importance for the accuracy of the results. Unfortunately, these parameters are often not mentioned in publications, and thus it is impossible to check the published results. Therefore, the numerical parameters of this work are enlisted and briefly explained in this chapter.

B.1 Parameters set in ABAQUS

All FE analyses in this work are carried out with ABAQUS. The default set of parameters to control the convergence of the non-linear solution technique (cp. 3.1.3) is slightly adopted. First, the ABAQUS keycard

```
*CONTROLS,ANALYSIS=DISCONTINUOUS
```

is used, to set the parameters $I_0 = 8$ and $I_R = 10$. I_0 is the number of equilibrium iterations after which the check is made that the residuals are not increasing in both two consecutive iterations. I_R is the number of equlibrim iterations after which the logarithmic rate of convergence check begins. Both are set to avoid premature cutbacks of the time increment in difficult analyses.

The keycard

```
*CONTROLS,PARAMETERS=FIELD,FIELD=DISPLACEMENT
0.1,1.0,,
```

allows to set the tolerances for convergence control. The statement above sets the convergence criterion for the largest residual displacement ($= \max(\Delta q_i)$)

to the average displacement (u_{mean}) to $R_n^\alpha = 0.1$. The convergence criterion for the largest solution correction (Δq_{i+1}) to the corresponding incremental solution value (Δq_i) is set to $C_n^\alpha = 1.0$.

To improve the robustness of the Newton method, the line search algorithm is used. In initial iterations, where ABAQUS is comparatively far from the solution, residuals $\mathbf{r}$ may be large and so may be the correction $\Delta\mathbf{q}$. The line search algorithm scales the correction by a factor s^{ls} which is determined by the line search algorithm. The command below sets the maximum number of search iterations to $N^{ls} = 4$.

```
*CONTROLS, PARAMETERS=LINE SEARCH
4
```

B.2 Parameters set in UMAT

The time integration of state variables requires a second sub step loop beside the Newton-Raphson iteration. This brings along the need of additional criterias to keep the errors small resulting from numerical integration. The algorithm used in this work (see 3.1.4) yields an error estimation that is used to control sub step size (eq. 3.32 to eq. 3.34). To meet the requirements of accuracy and speed a parametric study was performed, where the increment size and the values of error controls TOL have been varied. A standard triaxial element test has been chosen to perform the study. As a result intergranular strains and stress rates must have both their own error controls (TOL_T and TOL_{IGS}) to provide accurate results in a tractable time. The parametric stud showed, that a combination of $TOL_T = 0.001$ with $TOL_S = 0.00001$ gives sufficent accurate results at a manageable calculation time.

Appendix C

Model tests using Soiltron

Model tests under normal graviational conditions have limited applicability due to the stress dependency of soil behaviour. To overcome this shortcoming Soiltron has been developed at the Institute of Geotechnical and Tunnelling Engineering of the University of Innsbruck [58]. The idea of Soiltron is to simulate the soil behaviour at field conditions (higher pressure due to greater depth) in a small scale model (lower pressure due to low depth) with a loose soil. This means, that the effects of barotropy (dependence of soil behaviour on stress) are compensated using the effect of pyknotropy (dependence of soil behaviour on density). The mechanical similarity is achieved by adding soft and light particles (polystyrene beads or perlite grains) to the prototype soil.

Within the scope of this thesis six model tests with soiltron have been conducted.

C.1 Material Parameters and initial density

Soiltron used in this work is a mixture of Ottendorf-Okrilla quartz sand (section 4.2.2) with perlite (see Fig. C.1) at a mass ratio of 1 g sand to 0.0134 g perlite.

Up to now two different types of Soiltron have been developed: Soiltron with polystyrene beads and Soiltron with perlite grains. When this test series was started, the segregation of sand and polystyrene beads was a yet unsolved problem, therefore Soiltron with perlite grains was adopted.[1]

Soiltran has been built in in layers. The material has been poured in layers of 5 cm with 25 cm height of fall. Then a rake was drawn through the layer to guarantee a uniform and reproducible state.

[1] In the meantime this problem is solved. Adding ethanol causes a slight cohesion and stops segregation. Ethanol then evaporates very fast from the mixture.

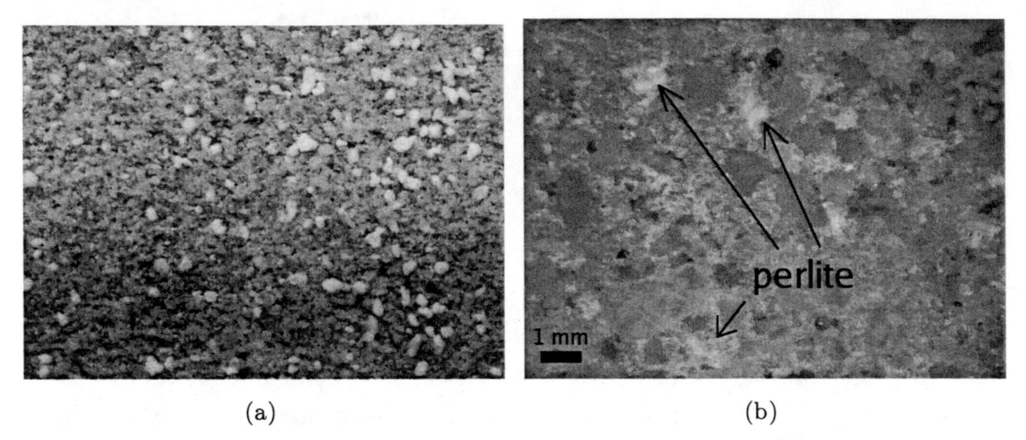

(a) (b)

Fig. C.1: Soiltron [58]: (a) Ottendorf-Okrilla sand and perlite grains; (b) Ottendorf-Okrilla sand and perlite 12-fold magnified.

Test No.	void ratio e	rel. density index I_e	pre-paration	material	change of diameter ΔD	in-crements
S0.66-1	0.65	0.32	raked	Soiltron	0.66	4
S0.66-2	0.65	0.30	raked	Soiltron	0.66	4
S0.66-3	0.65	0.31	raked	Soiltron	0.66	4
S1.0-1	0.72	0.08	tamped	Soiltron	1.00	6
S1.0-2	0.63	0.37	raked	Soiltron	1.00	6
S1.0-3	0.62	0.40	raked	Soiltron	1.00	6

Tab. C.1: Initial conditions of tests with Soiltron.

C.2 Results

In Fig. (C.3) and Fig. (C.4) the model tests using Soiltron are compared to the tests with Ottendorf-Okkrilla sand (chapter 4). Relative density index is plotted vs. normalised volume loss and normalised settlements, respectively. The outlier is due to segregation effects during build in of the first test with Soiltron (Fig. C.2a). In the following tests a rake was drawn through every layer and segregation was avoided successfully (Fig. C.2b).

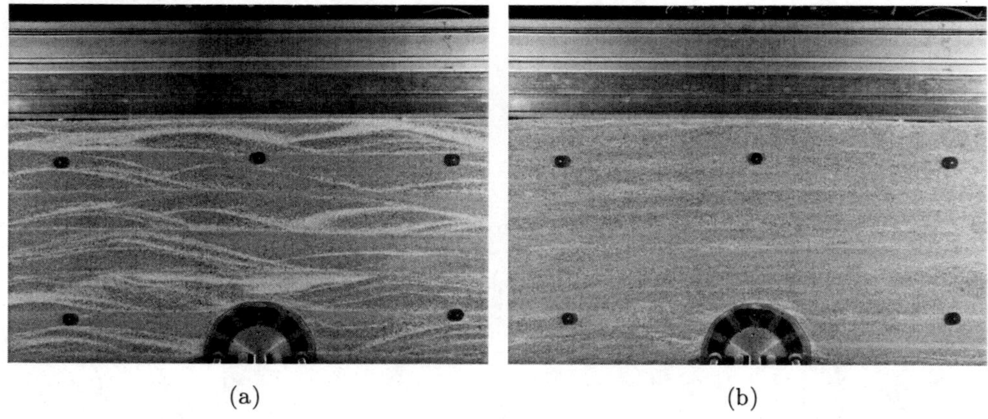

(a) (b)

Fig. C.2: (a) Segregation of perlite during build in; (b) Due to additional preparation with a rake segregation is remedied.

As previously mentioned, the idea of Soiltron is to produce a loose initial state of soil by adding soft additives in order to get the same soil behaviour as under higher (prototype) pressure. One of the assumptions of Soiltron is, that the additives are so soft, that they act as artificial pores. As a result, the settlements from tests with Soiltron should be located on the same line in the settlement vs. relative density diagram as the settlements from tests with Ottendorf-Okrilla sand do. In Fig. (C.5b) this assumption is disproved. Note, that the relative density in this diagram is calculated assuming that perlite is air. The void ratio then reads:

$$e = \frac{\gamma_s}{\gamma_{d,sand}} - 1 \quad , \tag{C.1}$$

with $\gamma_{d,sand} = \gamma_d / 1.01333$. The factor 1.0133 results from the ratio of masses 0.0133 g Perlite per 1 g sand.

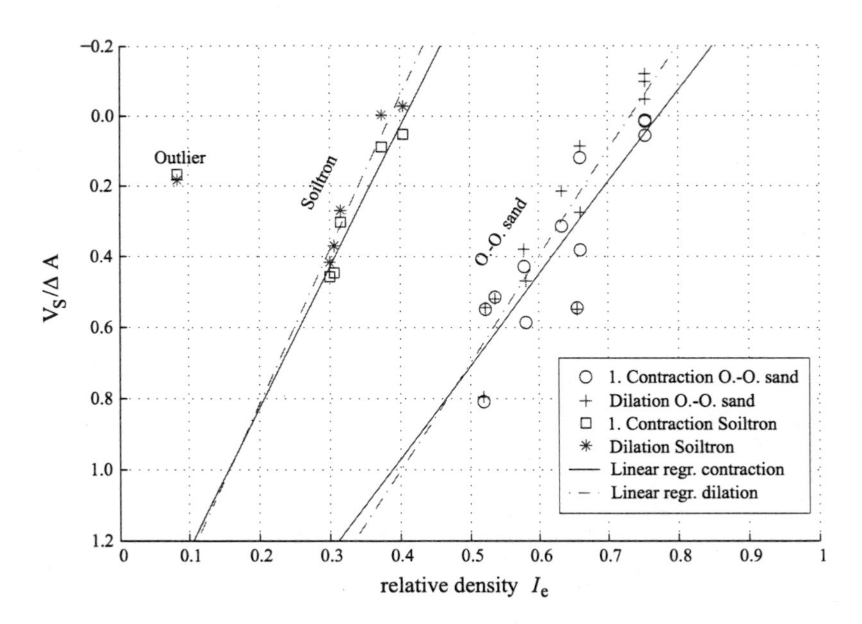

Fig. C.3: Volume of settlement trough V_S vs. relative density index I_e.

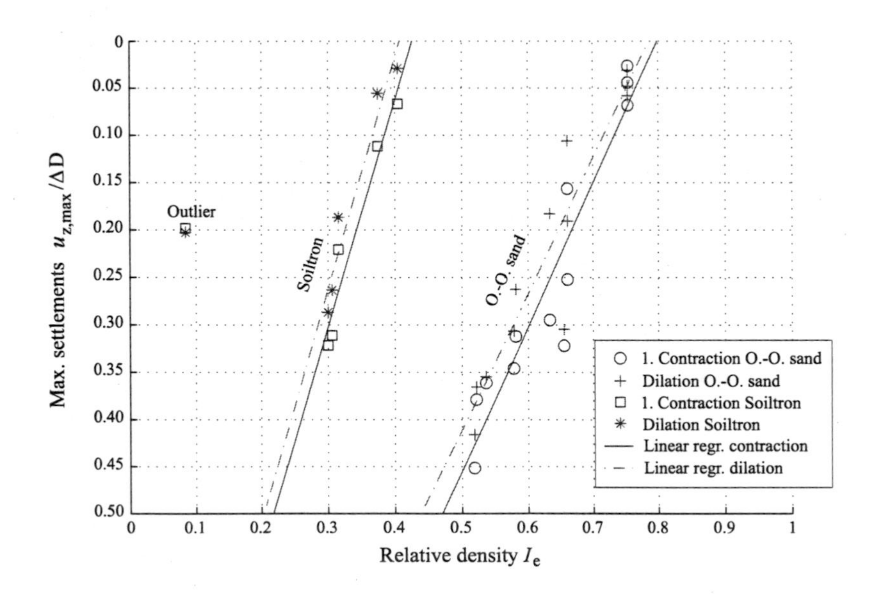

Fig. C.4: Max. settlements u_z vs. relative density index I_e.

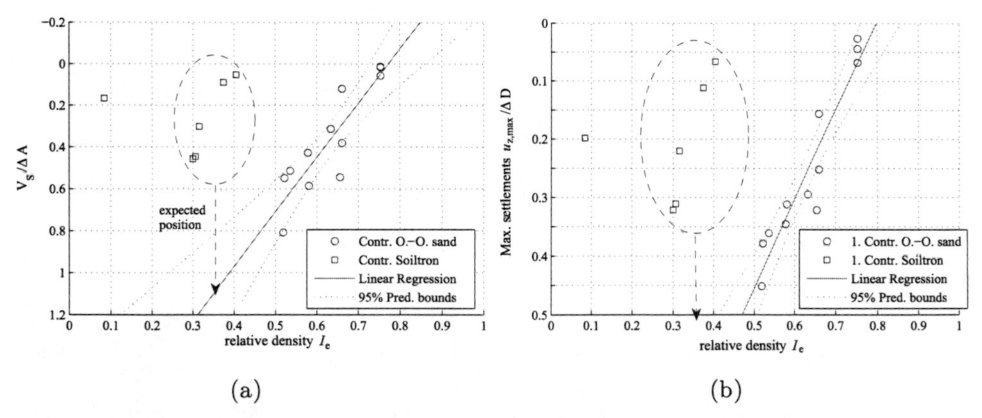

Fig. C.5: (a) Volume of settlement trough V_S vs. relative density; (b) max. settlements vs. relative density. Relative density calculated assuming perlite is pore volume.

The behaviour of Soiltron is much stiffer than expected. This is due to the small stresses in the model, which are in the range from $0 - 15\ \mathrm{kN/m^2}$. At the depth of tunnel axis the initial vertical stress of $0.425\ \mathrm{m} \cdot 17.5\ \mathrm{kN/m^3} \approx 7.5\ \mathrm{kN/m^2}$ can be assessed. Obviously, perlite is too stiff at these low stresses. Thus the perlite grains do not behave like air pores, but like sand grains.

Assuming the perlite grains behaves like sand grains leads to a void ratio of $e = 0.35$[2], which appears as too low. Therefore the relative density is calculated using $e_{\min}$ and $e_{\max}$ of Soiltron. However, using this assumption the results with Soiltron are still not in line with the ones with pure sand (Fig. C.6a and Fig. C.6b).

C.3 Summary

The presented tests with Soiltron are the first ones using this technique in model tests. Due to mixture with a cement mixer and the preparation of the sand with a rake, perlite grains may have been broken before the test started. Moreover Soiltron has been used for several tests, which is another cause of broken perlite grains. However, smaller grains of the same material are in general stiffer than larger ones. Thus, the too stiff behaviour of Soiltron may be caused by grain crushing previous to the model tests. Further tests with polystyrene are desirable to investigate the influence of stiffness of the additives on model tests with low stresses.

[2] Assuming the unit weight of perlite as $\rho_{\text{perlite}} = 1.5\ \mathrm{kN/m^3}$.

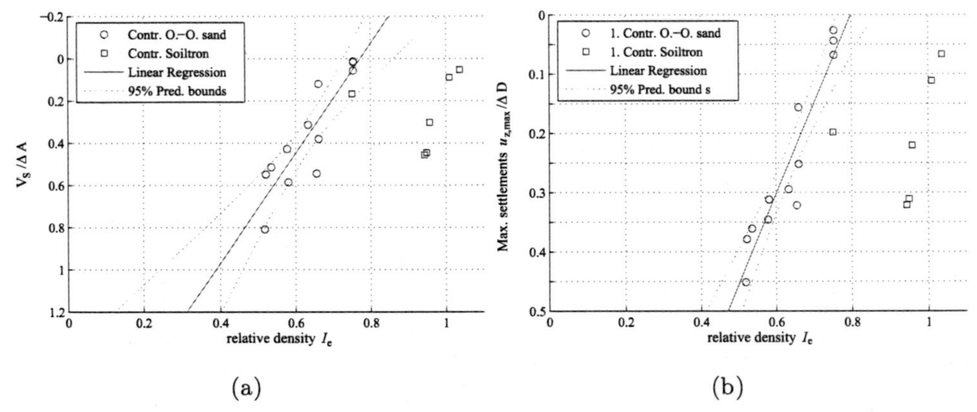

(a) (b)

Fig. C.6: (a) Volume of settlement trough V_S vs. relative density; (b) max. settlements vs. relative density. Relative density is calculated with $e_{\min}$ and $e_{\max}$ from Soiltron.